MANUAL FOR
MOS USERS

MANUAL FOR
MOS USERS

John D. Lenk

Consulting Technical Writer

Reston Publishing Company, Inc.
A Prentice-Hall Company

Reston, Virginia

Library of Congress Cataloging in Publication Data

Lenk, John D
  Manual for MOS users.

  1. Metal oxide semiconductors. 2. Electronic
apparatus and appliances. I. Title.
TK7871.85.L448   621.3815'2   74–23929
ISBN 0–87909-478-8

© 1975
Reston Publishing Company, Inc.
A Prentice-Hall Company
Box 547
Reston, Virginia 22090

10 9 8 7 6 5 4 3 2 1

Printed in the United States of America.

Dedicated to my wife

Irene

This is her third sea-house book

CONTENTS

PREFACE

As the title implies, this manual is written for *users* of MOS (Metal Oxide Semiconductor) devices, rather than for designers of MOS devices. The manual is, therefore, written on the basis of using existing, commercial MOS devices to solve design and application problems.

Typical users include design specialists who want to incorporate MOS devices in their electronic units and systems, or technicians who must service, test, and troubleshoot equipment containing MOS devices. Other groups that might make good use of this data are experimenters and hobbyists.

The approach used serves a two-fold purpose: (1) it acquaints the readers with MOS devices, in general, so that the users can select commercial units to meet their particular circuit requirements, and (2) it shows the readers the many applications for existing MOS units, not always apparent from the data in manufacturer's catalogs.

The manual assumes that the reader is already familiar with basic electronics, including discrete and IC forms, but has little or no knowledge of MOS principles or devices. For this reason, Chap. 1 provides an introduction to MOS. Such topics as the how and why of MOS fabrication techniques, a comparison of MOS devices to comparable units using other fabrication methods, a description of basic physical types and circuit types found in commercial MOS units, as well as some basic design considerations are included. Chap. 1 also discusses practical considerations for all types of MOS devices.

As the reader will soon discover, the two most common types of MOS are the discrete MOSFET, and the complementary (COS or CMOS) integrated circuit (IC). The uses for discrete MOSFETs are discussed in Chap. 2 through 4. Complementary MOS units are covered in Chap. 5.

ix

Throughout all of these chapters, considerable emphasis is placed on interpreting manufacturer's datasheets and other literature.

Often, experimenters must work with MOS units for which complete characteristic data is not available. Under these circumstances, it is necessary to test the device using simulated operating conditions. For this reason, typical test data is included in Chap. 6.

As with any solid-state device, it is possible to apply certain approximations or guidelines for the selection of circuit component values. These rules-of-thumb can be stated in basic equations, requiring only simple arithmetic for the solutions. This manual starts with rules-of-thumb for the selection of circuit components, on a trial-value basis, assuming a specified goal and a given set of conditions. The manual concentrates on simple, practical approaches to MOS use, not on MOS analysis. Theory is kept at a minimum.

The values of circuit components used with MOS devices depend on device characteristics, available power sources, the desired performance (voltage amplification, stability, etc.), and existing circuit conditions (input/output impedances, signal amplitudes, etc.). The MOS characteristics are to be found in the manufacturer's data. The overall circuit characteristics can then be determined, based on a reasonable expectation of the MOS characteristics. Often, the final circuit is a result of many trade-offs between desired performance and available characteristics. This manual discusses the problems of trade-off from a simplified, practical standpoint.

Since the manual does not require advanced mathematics or theoretical study, it is ideal for the experimenter. On the other hand, the manual is suited to schools where the basic teaching approach is circuit analysis, and where a great desire exists for practical design.

The author has received much help from many organizations and individuals prominent in the field of MOS technology. He wishes to thank them all, particularly Motorola Semiconductor Products, Inc., The Solid State Division of Radio Corporation of America (RCA), and the Semiconductor Group of Texas Instruments Incorporated.

The author also wishes to express his appreciation to Mr. Joseph A. Labok of Los Angeles Valley College for this help and encouragement.

John D. Lenk

1. INTRODUCTION TO MOS DEVICES

The MOS (metal oxide semiconductor) principle can be used to fabricate almost any solid-state device, either discrete or IC (integrated circuit). However, the MOS principle is, at present, mostly used to produce some form of FET (field effect transistor). In discrete form, the most common device is the MOSFET (also known as the IGFET, or insulated gate field effect transistor). The most frequent use of MOS in integrated circuits is the *complementary* MOS (generally known as CMOS, COS/MOS, or some similar term, depending on the manufacturer). This chapter concentrates on the basics for both of these MOS types.

1–1. BASIC MOS CONSTRUCTION AND THEORY

Figure 1–1 shows the development of an *N*-channel MOSFET. The basic MOSFET is essentially a bar of doped silicon, or some similar *substrate* material, that acts like a resistor. The terminal into which current is injected is called the *source*. The source terminal is similar in function to the cathode of a vacuum tube. The opposite terminal is called the *drain* terminal, and can be likened to a vacuum tube plate. However, in a MOSFET, the polarity of the voltage applied to the drain and source can be interchanged.

In Fig. 1–1(a), the substrate is a high resistance *P*-type material. Two separate low resistance *N*-type regions (source and drain) are diffused into the substrate, as shown in Fig. 1–1(b). The surface of the structure is covered with an insulating *oxide* layer, illustrated in Fig. 1–1(c). Holes are cut into the oxide layer, allowing *metallic* contact to the source and

1

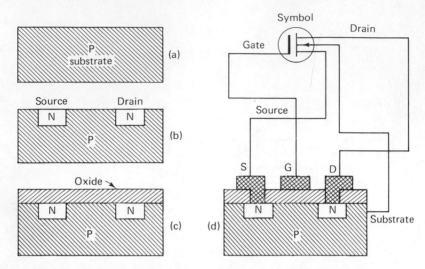

Fig. 1–1 Typical *N*–channel MOSFET construction and symbol.

drain. The *gate* metal area is overlayed on the oxide, covering the entire channel region. Similar metal contacts are made to the drain and
source, as shown in Fig. 1–1(d). The contact to the metal area covering
the channel is the *gate* terminal. There is no physical penetration of the
metal through the oxide into the substrate. Since the drain and source
are isolated by the substrate bulk, any drain-to-source current that occurs
in the absence of gate voltage is very low.

The metal area of the gate, in conjunction with the insulating oxide
layer and the semiconductor channel, forms a *capacitor*. The metal area
is the top plate; the substrate (bulk) material is the bottom plate.

Figure 1–2 shows operation of the MOSFET. Positive charges at the
metal side of the metal-oxide *capacitor* induce a corresponding negative charge at the semiconductor side. As the positive charge at the gate
is increased, the negative charge *induced* in the semiconductor increases
until the region beneath the oxide becomes an *N*-type semiconductor
region and current can flow between the source and drain through the
induced channel. Drain current flow is *enhanced* by the gate voltage
and can be controlled or modulated by it. Channel resistance is directly
related to the gate voltage.

Note that it is possible to make a MOSFET with a *P*-channel by re-
versing all of the material types. Likewise, it is possible to form both
N-channel and *P*-channel MOSFETs on the same substrate. This results in the complementary COS/MOS or CMOS types used in digital
circuits. The fabrication and theory of complementary MOS devices is
discussed in Sec. 1–3.

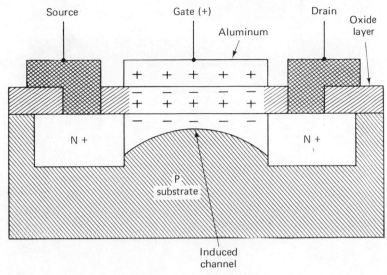

Fig. 1–2 Operation of typical MOSFET showing channel enhancement.

The MOSFET just described is called an *enhancement* type. A *depletion* type MOSFET is formed when an *N*-type channel is diffused between the source and drain so that drain current can flow when the gate voltage is zero. Such an arrangement is shown in Fig. 1–3. The structure shown in Fig. 1–3(a) is both an enhancement and a depletion MOSFET.

When positive gate voltages are applied, the structure enhances in the same manner as in Fig. 1–2 (increased negative in the induced *N*-type channel between source and drain). When negative gate voltages are applied, the channel begins to deplete carriers as depicted in Fig. 1–3(b) (decreased negative, or increased positive, in the induced *N*-type channel between source and drain).

Note that Fig. 1–3(b) shows a greater number of positive carriers (or less negative carriers) nearest the drain terminal. This is because in any depletion-type FET drain current flow depletes the channel area nearest the drain terminal first.

Typical MOSFET symbols for both *P*-channel and *N*-channel are shown in Fig. 1–4. Note that the substrate (or bulk) arrow points away from the substrate in a *P*-channel device. This appears to be inconsistent with two-junction (bipolar) transistors where arrows point toward the substrate in PNP devices. However, it will be seen that the symbols are consistent when it is realized that *P*-channel MOSFETS are formed of *N*-type substrate material (and *N*-channel MOSFETS use *P*-type material).

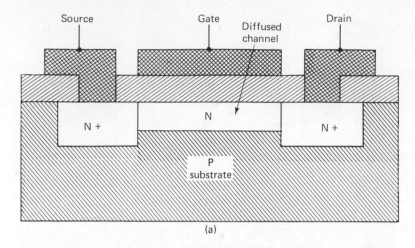

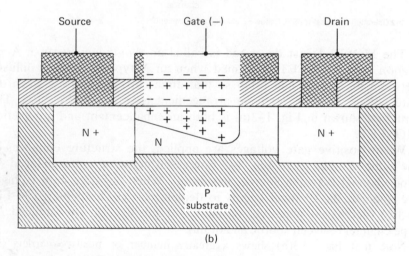

Fig. 1–3 Operation of typical depletion-type MOSFET showing channel depletion.

1–1.1 Operating Modes of MOS Devices

MOSFETs can be operated in three modes: *depletion only, enhancement only,* and a combination of *enhancement and depletion.*

The basic differences between these three modes can most easily be understood by examining the transfer characteristics of Fig. 1–5.

The *depletion only* MOSFET, classified as Type A, has considerable drain current flow for zero gate voltage. No forward gate voltage is re-

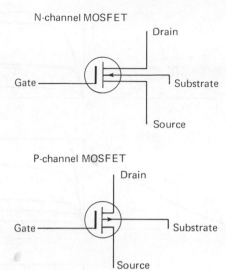

Fig. 1–4 Typical P–channel and N–channel MOSFET symbols.

quired. Instead, drain current is reduced by applying a reverse volt-age to the gate terminal. In the case of an *N*–channel MOSFET, a negative bias is required on the gate (between gate and source) to main-tain the drain current at some fixed level (or at zero). Positive and negative signal voltages are applied to vary the drain current above and below the level fixed by the gate-source bias.

The *depletion/enhancement* MOSFET, classified as Type *B*, also has considerable drain current flow for zero gate voltage (but not as much as *depletion only*). Drain current is increased by forward gate voltage and decreased by reverse gate voltage, permitting a wide variation in drain current. This represents one of the advantages of MOS devices using the insulated gate principle. With junction devices, such as bipolar transistors and JFETs (junction FETs), drain current can be increased by control voltage only until the PN junction becomes forward biased. At this point, a further increase in forward control voltage will not pro-duce an increase in current (saturation occurs).

The *enhancement only* MOSFET, classified as Type *C*, has little or no current flow for zero gate voltage. Drain current does not occur un-til a forward gate voltage is applied. This voltage is known as the *thres-hold* voltage, and is indicated in Fig. 1–5 as $V_{GS(th)}$. Once the thres-hold voltage is reached, the transfer characteristics for a Type *C* MOSFET are similar to those of a Type *B*.

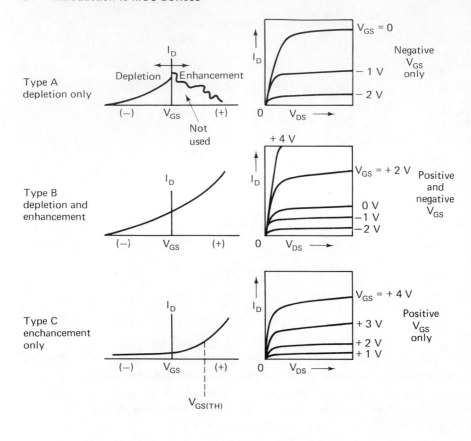

I_D = drain current   V_DS = drain-source voltage   V_GS = gate-source voltage

Fig. 1–5 Basic operating modes of MOSFETs. Device shown is *N*–channel. Reverse polarities for *P*–channel.
I_D = Drain current; V_{DS} = Drain-source voltage; V_{GS} = Gate-source voltage.

1–1.2 Equivalent Circuit of a MOSFET

An equivalent circuit for a typical MOSFET is illustrated in Fig. 1–6. The circuit described is for an enhancement (Type *C*) device. The symbol $C_{g(ch)}$ is the distributed gate-to-channel capacitance, representing the oxide capacitance. C_{gs} is the gate-source capacitance of the metal gate area that overlaps the source. C_{gd} is the corresponding gate-drain capacitance of the metal gate area overlapping the drain. $C_{d(sub)}$ and $C_{s(sub)}$ are junction capacitances from drain to substrate, and source to substrate. Y_{fs} is the transadmittance between drain current and gate source voltage. The modulated channel resistance is r_{ds}. R_D and R_S are the bulk resistance of the drain and source. The input resistance of the MOSFET is exceptionally high because

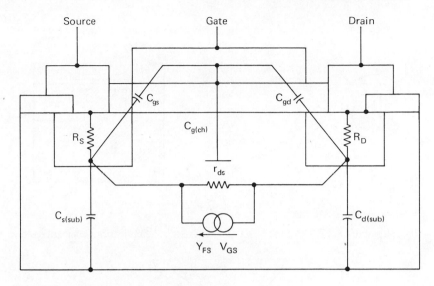

Source   Gate   Drain

Fig. 1–6 Equivalent circuit of enhancement mode MOSFET.

the gate acts as a capacitor with very low leakage. A typical input resistance is 10^{14} ohms. The output impedance is a function of r_{ds} (which is related to gate voltage) and the drain and source bulk resistances (R_D and R_S).

To turn the MOSFET on, the gate-channel capacitance $C_{g(ch)}$, the Miller capacitance C_{gd}, and the drain-substrate capacitance $C_{d(sub)}$ must be charged. The resistance of the substrate determines the peak charging current of $C_{d(sub)}$.

1–2. PRACTICAL MOSFET DEVICES

Typically, discrete MOSFETs are available in *single-gate* and *dual-gate configurations.*

Figure 1–7 shows the symbol, fabrication schematic, and chip layout of a typical single-gate MOSFET. The device shown is representative of single-gate MOSFETs manufactured by RCA. Such devices are contained in a standard TO–72 metal package, the outlines of which are shown in Fig. 1–8. Note that there are four terminal leads. These accommodate the drain, source, gate and bulk or substrate. Not all MOSFETs are provided with a lead to the substrate. In some cases, the substrate is connected to the source by means of an internal lead. In the MOSFET of Figs. 1–7 and 1–8, the fourth lead provides for electrical connection to the substrate and the metal case.

Figure 1–9 illustrates the fabrication schematic of a device representative of the MOSFETs manufactured by Motorola. The fabrication

1. Drain
2. Source
3. Gate
4. Bulk
   (substrate and case)

Symbol

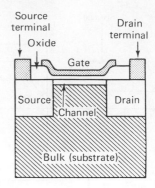

Fabrication schematic

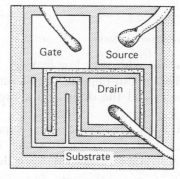

Chip Layout

Fig. 1–7 Symbol, fabrication schematic, and chip layout of a typical single-gate MOSFET. (Courtesy RCA)

is similar to that for typical MOS devices, except that a silicon-nitride passivation process is used. That is, the surface of the substrate is covered with the conventional insulating oxide layer, plus a nitride layer. The oxide layer serves as a protective coating, and to insulate the channel from the gate. However, the oxide is subject to contamination by sodium ions which are found in varying quantities in all environments. Such contamination results in long term instability and changes in device characteristics. Silicon nitride is impervious to sodium ions, and thus is used to shield the oxide layer from contamination. Holes are cut into both the oxide and nitride layers allowing metallic contact to the source and drain in the normal manner. Of course, there is no physical penetration of the gate metal through the oxide and nitride layers to the substrate. Note that the silicon-nitride passivation process also reduces susceptibility to damage from static-charge buildup

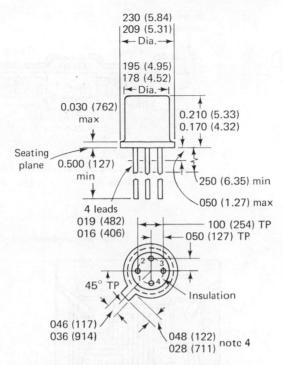

Fig. 1–8 Dimensional outline for standard JEDEC TO–72 package. (Courtesy RCA)

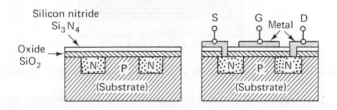

Fig. 1–9 Fabrication schematic of MOSFET with silicon-nitride passivation. (Courtesy Motorola)

during handling of the MOSFET. The subject of static-charge buildup is discussed further in Secs. 1–4 and 1–5.

Figure 1–10 shows the symbol, fabrication schematic, and chip layout of a typical dual-gate MOSFET. The device shown is representative of dual-gate MOSFETs manufactured by RCA, and features a series arrangement of two separate channels. Each channel has independent control. This arrangement results in substantially lower feedback capacitance, greater gain, remote AGC capability in RF amplifier applications, substantially better cross-modulation characteristics and lower spurious response than are provided by single-gate types. (The characteristics and advantages of MOSFET devices are discussed

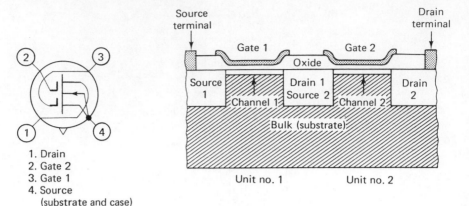

1. Drain
2. Gate 2
3. Gate 1
4. Source
   (substrate and case)

Symbol   Fabrication schematic

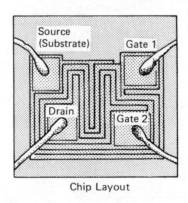

Chip Layout

Fig. 1–10 Symbol, fabrication schematic, and chip layout of a typical dual-gate MOSFET. (Courtesy RCA)

in Secs. 1–6 and 1–7.) The availability of two independent control gates also offers unique advantages for chopper, clipper and gated-amplifier service, and for applications involving the combination of two or more signals, such as mixers, product detectors, color demodulators, and balanced modulators.

Note that the dual-gate device is also contained in a standard TO–72 metal package with four terminal leads. Three of the leads accommodate the drain; gate 1, and gate 2, respectively. The fourth lead provides electrical connection to the source, substrate, and metal case.

With the series dual-gate MOSFET, the drain current is controlled by voltages on both gates, since current from source to drain must pass through both channels. There is a common region between channel 1 and channel 2 that acts as a drain for channel 1 and a source for channel 2. This region does not have a terminal lead, and is not accessible from outside the package (case).

1–2.1 Transfer Characteristics for Basic MOSFETs

Figures 1–11 and 1–12 show the transfer characteristics for typical single-gate and dual-gate MOSFETs, respectively. The transfer characteristic shown in Fig. 1–11 is for an *N*-channel, depletion-type, single-gate MOSFET. Note that there is heavy drain current flow (about 15 mA) at zero gate voltage. When the gate voltage is at about −2.5V, the drain current drops to zero.

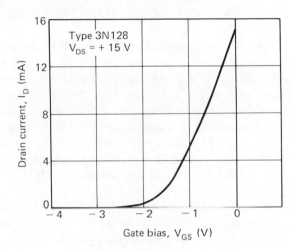

Fig. 1–11 Transfer characteristic for typical single-gate MOSFET. (Courtesy RCA)

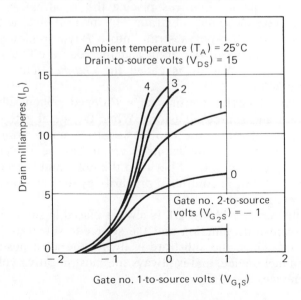

Fig. 1–12 Transfer characteristic for typical dual-gate MOSFET. (Courtesy RCA)

The transfer characteristic shown in Fig. 1–12 is for an *N*-channel, depletion-type, dual-gate MOSFET. Note that the drain current is controlled by voltages on *both* gates. For example, with zero volts on gate 2, the drain current can not exceed about 7 mA, even with gate 1 at +1V. If gate 2 is reduced to −1V, then the drain current cannot exceed about 2 mA. If gate 2 is raised to +2V, then the drain current can be increased about 14 mA, with gate 1 at (or near) zero.

1–3. PRACTICAL COMPLEMENTARY MOS DEVICES

Typically, integrated circuit (IC) MOS devices are available in complementary form. These devices are primarily for use in digital circuits, and consist of a combination of *P*-channel and *N*-channel units on a common chip.

The basic physical construction, and corresponding diagrams, for both *P*- and *N*-channels of a complementary MOS integrated circuit are shown in Fig. 1–13. The device illustrated is representative of complementary MOS IC devices manufactured by Motorola. The following is a summary of *P*- and *N*-channel operation, and MOSFET bias relationships, as they relate to digital ICs.

The *P*-channel device consists of a lightly *N*-doped silicon substrate with heavily-doped *P*-type diffusions into this substrate. Between the drain and source is the gate-oxide region, which serves as the insulation between the metal gate and the substrate. The basic operation of the *P*-channel device involves placing the metal gate at a negative potential with respect to the substrate. The induced electric field causes an inversion of the *N*-type substrate into a *P*-type region. This inversion occurs only between the drain and the source diffusions. The inverted area of the substrate is called the *channel*. The carriers in a *P*-channel are *holes*.

The *N*-channel device consists of a *P*-doped silicon substrate with *N*-doped drain and source diffusions. When the metal gate is placed at a positive potential with respect to the substrate, an *electron-dominated* channel between the two diffusions is created in the *P*-type substrate, resulting in the flow of current between the drain and source. The magnitude of current flow is controlled primarily by the gate-to-substrate potential difference, or bias.

The complementary MOSFET is always biased in such a manner that the drain-to-substrate junction and the source-to-substrate junction are *reverse-biased*. Thus, the substrate is always the most positive voltage on the *P*-channel device, and is always the most negative voltage on the *N*-channel device.

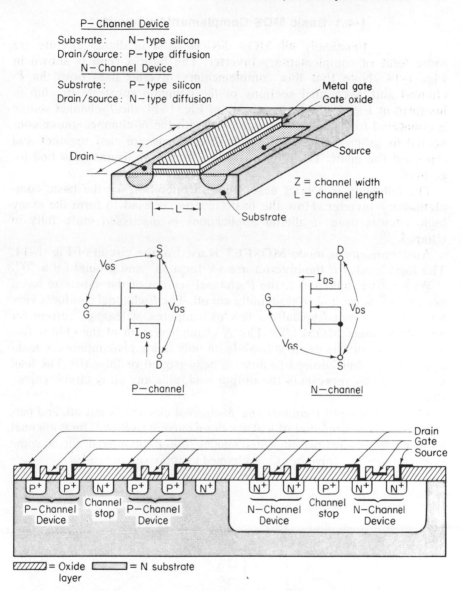

P– Channel Device

Substrate :   N–type silicon
Drain/source: P–type diffusion

N– Channel Device

Substrate: P– type silicon
Drain/source: N– type diffusion

Z = channel width
L = channel length

P–channel   N–channel

= Oxide   = N substrate
layer

Fig. 1–13 Basic physical construction and diagrams for P– and N–channel CMOS IC.

Physically, the drain diffusion is identical to the source diffusion. The two diffusions, however, are usually distinguished when the device is used in a circuit. The diffusion at the *least potential difference* with respect to its substrate is called the *source*.

1–3.1 Basic MOS Complementary Inverter

Practically all MOS devices used in digital circuits are some form of complementary inverter. The basic circuit is shown in Fig. 1–14. Note that this complementary inverter uses both the *P*-channel and *N*-channel sections of the common substrate or chip illustrated in Fig. 1–13. In the circuit of Fig. 1–14, the *P*-channel source is connected to the supply voltage (+V) with the *N*-channel source connected to ground. The gates of both channels are tied together and represent the input. The output is taken from both drains (also tied together).

The following is a brief description of operation for the basic complementary inverter. How the basic inverter is used to form the many logic circuits used in digital applications is discussed more fully in Chap. 5.

An enhancement mode MOSFET is used in the circuit of Fig. 1–14. The logic levels for the inverter are +V for a "1" and ground for a "0".

With a true input (+V), the *P*-channel section of the substrate has a zero gate voltage and is essentially cut off. The *P*-channel conducts very little drain current (typically a few picoamperes of leakage current for an enhancement MOSFET). The *N*-channel section of the chip is forward-biased and its drain voltage (with only a few picoamperes of leakage drain current allowed to flow) is near ground or false (0). The load capacitance C_L represents the output load, plus any stray circuit capacitance.

With a false input (ground), the *N*-channel element is cut off, and permits only a small amount of leakage drain current to flow. The *P*-channel element is forward-biased, thus making the *P*-channel drain at some voltage near +V. Capacitor C_L is charged to approximately +V.

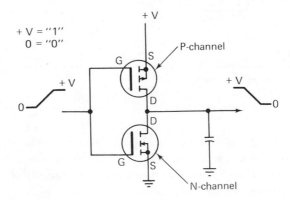

Fig. 1–14 Basic MOS complementary inverter.

No matter which logic signal is applied at the input (or appears at the output), power dissipation is extremely low. This is because both stable states (true and false or 1 and 0), are conducting only a few picoamperes of leakage current (since both channels are in series, and one channel is always cut off). Power is dissipated only during switching, an ideal situation for logic circuits.

In addition to the lower power dissipation, another advantage of MOS devices for digital logic circuits is that no coupling elements are required (the gate acts as a coupling capacitor). Since no capacitors are needed, it is relatively simple to fabricate MOSFET logic elements in IC form (fabrication of capacitors is usually a major stumbling block for most IC design). OR, NOR, AND and NAND gates, with either positive or negative logic, can be implemented with MOSFETs. Thus, almost any logic circuit combination can be produced using the basic MOS complementary inverter. MOSFET logic can also be used over a wide range of power supply voltages.

1-4. HANDLING MOS DEVICES

Electrostatic discharges can occur when a MOS device is picked up by its case and the handler's body capacitance is discharged to ground through the series arrangement of the bulk-to-channel and channel-to-gate capacitances of the device. This applies to both discrete MOSFETs and complementary MOS ICs. Although MOS devices, like some high-frequency bipolar transistors, are susceptible to damage by the electrostatic discharge of energy through the devices, many MOS devices are currently being used extensively in all phases of electronics without damage. Of course, this requires proper handling, particularly when the MOS device is out of the circuit. In circuit, a MOS device is just as rugged as any other solid state component of similar construction and size.

MOS devices are generally shipped with the leads all shorted together to prevent damage in shipping and handling (there will be no static discharge between leads). Figure 1-15 shows a typical MOSFET with a shorting spring installed. This shorting spring, or similar device, *should not be removed until after* the device is soldered into the circuit. An alternate method for shipping or storing MOS devices is to apply a conductive foam between the leads. Such a foam is manufactured by Emerson & Cuming, Inc., listed as ECCOSORB LS26.

Note that polystyrene insulating "snow" is not recommended for shipment or storage of MOS devices. Such snow can acquire high static charges which could discharge through the device.

If it becomes necessary to install any MOS device in a circuit, or to remove the device from the circuit for test or some other operation, consider the following points.

1. When removing or installing a MOS device, first turn the power off. If the MOS device is to be removed, your body should be at the *same potential* as the unit from which the device is removed. This can be done by placing one hand on the chassis *before* removing the device. If the MOS device is to be connected to an external circuit (for test, etc.) put the hand holding the MOS device against the tester front panel and connect the leads from the tester to the device leads. This procedure prevents possible damage from static charges on the tester (or other external unit).

   WARNING: Make certain that the chassis or external unit is at ground potential before touching it. In some obsolete or defective equipment, the chassis or panel is above ground (typically by one-half of the line voltage).

2. When handling a MOS device, the leads must be shorted together. Generally, this is done in shipment by a shorting ring or spring (Fig. 1–15). When testing MOS devices, connect the tester lead to the MOS (preferably the source lead first), then remove the shorting ring.

3. When soldering or unsoldering a MOS device, the soldering tool tip must be at ground potential (no static charge). Connect a clip lead from the barrel of the soldering tool to the chassis or tester case. The use of a soldering gun is *not* recommended for MOS devices.

4. Remove power to the circuit before inserting or removing a MOS device (or a plug-in module containing a MOS device). The voltage transients developed when terminals are separated may damage the MOS unit. This same caution applies to circuits with conventional transistors. However, the chances of damage are greater with MOS devices.

Drawing shows shorting spring
which should not be removed
until after device is soldered
into circuit.

Fig. 1–15 Typical MOSFET with shorting spring installed on leads to prevent damage due to static discharge. (Courtesy RCA)

1–5. PROTECTING MOS DEVICES

Because of the static discharge problem described in Sec. 1–4, manufacturers provide some form of protection for a number of their MOS devices. Generally, this protection takes the form of a diode incorporated as part of the substrate material. This section describes such protective diodes, as found in certain RCA and Motorola MOS devices.

Before going into specific protection methods, let us analyze the breakdown mechanism associated with gate destruction. Figure 1–16 shows a typical single-gate MOSFET, including the structure [Fig. 1–16(a)], the symbol [Fig. 1–16(b)] and an over-simplified equivalent circuit [Fig. 1–16(c)]. A study of Fig. 1–16 will show the possible discharge paths within the device. Note that a substrate *diode* is formed by the PN junction integrated over the entire junction area (that is, the source and drain diffusions connected by the inversion layer or *N*-type channel). Figure 1–16(c) lumps the substrate-to-channel diode into one equivalent diode D_1 which terminates at the center of the channel.

C_{IN} in Fig. 1–16(c) represents the gate-to-channel capacitance. R_1 and R_2 represent the channel resistance. R_3 is the leakage resistance associated with the substrate-to-channel equivalent diode D_1. Leakage resistance across C_{IN} is intentionally deleted because such resistance is many thousands of times greater than R_3.

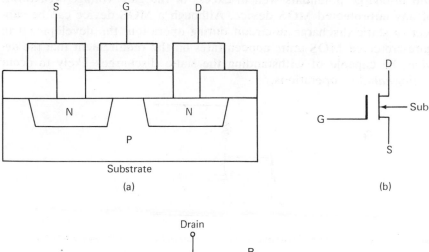

(a)

(b)

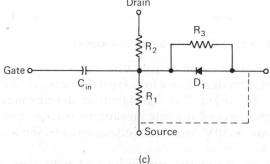

(c)

Fig. 1–16 Structure, symbol, and simplified equivalent circuit of typical single-gate MOSFET.

In a typical RCA MOSFET (the 3N128), C_{IN} is less than 5 pF. The channel resistance $R_1 + R_2$, which is a function of the applied bias, can range from 10^2 to 10^{10}. R_3 is also subject to variations determined by operating conditions, but can be assumed to be in the order of 10^9 ohms. Thus, a dc voltage applied between the gate and any other element results in practically all of this voltage being applied across C_{IN}.

Typically, the gate voltage-handling capability of a MOS device (without protection) is not over 30V, although some MOS devices can withstand gate-to-substrate voltages of about 100V and not result in breakdown. However, with any MOS device, once the oxide insulation breaks down, the device is destroyed.

Figure 1–17 depicts a simple equivalent circuit of a static discharge generator as it appears at the input of a MOS device. E_S represents the static potential stored in the static generator capacitor C_D. This voltage must be discharged through internal generator resistance R_S. As a point of reference, a human body acts as a static (storage) generator with a capacitance C_D ranging from 100 to 200 pF, and resistance R_S greater than 1000 ohms. The important point is that the human body can store and discharge potentials well in excess of the gate voltage-breakdown of any unprotected MOS device. Although a MOS device can be subject to static discharge in-circuit during operation, the development of gate-protected MOS units concentrates on the requirement that the devices be capable of withstanding the static discharges likely to occur during *handling* operations.

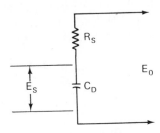

Fig. 1–17 Simple equivalent circuit of a static discharge generator as it appears at the input of a MOS device.

1–5.1 Basic Gate Diode Protection

The ideal situation in gate protection is to provide a signal-limiting configuration that allows for a typical sine wave to be handled without clipping or distortion. The signal-limiting devices must limit all transient voltages that exceed the gate breakdown voltage. For example, if the gate breakdown is 30V, transient voltage must be limited to something less (e.g. 25V).

One possible means of securing proper limiting is to place a diode in parallel with C_{IN} as shown in Fig. 1–18. This method is popular with some single-gate MOS devices, both discrete and IC, but has certain limitations. In terms of signal handling, the single diode of Fig. 1–18 clips the positive peaks of a sine wave when the device is operated near zero bias. The single diode method is not suitable for dual-gate MOS devices where one gate is operated with a fixed dc bias. For example, a dual-gate MOSFET is frequently operated with an RF signal superimposed on a slightly positive *bias* on one gate, and an AGC voltage on the other gate. The AGC voltage is capable of both positive and negative swings.

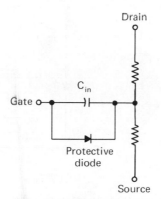

Fig. 1–18 Static discharge protection provided by single gate diode.

1–5.2 Back-to-Back Diode Protection

Whatever limiting device is used, it is important that the limiting device be an effective *open circuit* to any incoming signals, throughout the amplitude range of such signals. One of the best available methods for accomplishing this effective open circuit is the back-to-back diode arrangement (pioneered by RCA) shown in Figs. 1–19 and 1–20. The back-to-back method is generally used with dual-gate

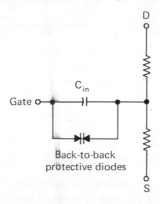

Fig. 1–19 Static discharge protection provided by back-to-back gate diodes

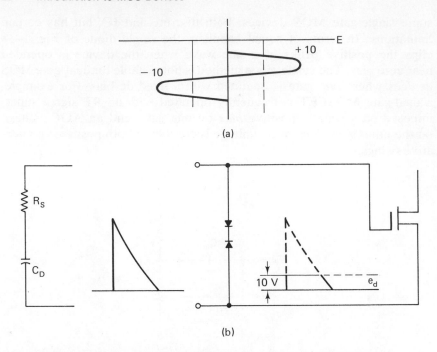

Fig. 1–20 Transfer characteristic of protective diodes and resulting waveforms in equivalent circuit.

MOS devices, and each gate is provided with a separate pair of back-to-back diodes.

Ideally, the transfer characteristic of the protective signal-limiting diodes has an infinite slope at limiting, as illustrated in Fig. 1–20(a). Under these conditions, the static potential generator in Fig. 1–20(b) discharges through the internal impedance R_S into the load represented by the signal-limiting diodes. The ideal signal-limiting diodes, with an infinite transfer slope ($R_S = 0$) will then limit the voltage (at the gates) to the value e_d. The difference voltage $E - e_d = e_s$ (where E is the static potential in the static generator, e_d is the diode voltage drop, and e_s is the voltage drop across the generator internal resistance) appears as an IR drop across R_S. The instantaneous value of the diode current is then equal to e_s/R_S. During handling of MOS devices, the practical range of this discharge varies from several milliamperes to several hundred milliamperes.

1–5.3 Practical Monolithic
Gate-Protected MOS Devices

Optimum protection is afforded to an MOS gate with signal-limiting diodes that exhibit zero resistance (that is, an infinite transfer slope and fast turn-on time) to all high-level transients. In addition, the diodes should ideally add no capacitance or loading to an RF input circuit. When these conditions are met, the high-frequency performance of the device will be comparable to that of an unprotected device. To keep cost as size at a minimum, it is best if the gate protection is accomplished by monolith methods.

One approach to integrating protective diodes into an MOS device is shown in Fig. 1–21. With such a scheme, the diodes provide adequate protection against the transients experienced in handling, in addition to those found when the device is installed in some piece of equipment. In this approach, the silicon substrate required for an N-channel, depletion-type MOS device is the starting material. The N-type wells are diffused into the silicon to provide pockets for the protective devices. The surface concentrations and the depths of these wells are carefully controlled because both factors are important in determining diode characteristics.

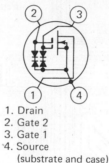

1. Drain
2. Gate 2
3. Gate 1
4. Source
   (substrate and case)

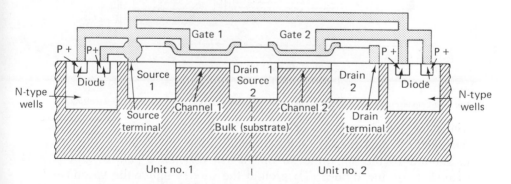

Fig. 1–21 Practical monolithic gate-protected MOS device. (Courtesy RCA)

The *P*+ regions are diffused into the *N*-type wells to form the diodes. *P*+ regions are also diffused around the periphery to isolate the diode structure from the surface of the MOS device, and to provide a region into which the channels may be terminated. The size of the diodes is determined by the desired current-handling capability, and by the amount of capacitance that can be tolerated across the gate of the MOS device. The spacing of the diodes is determined by the area available and the desired amount of transistor control action from diode to diode. After the diode structures are formed, they are covered by a protective oxide. The MOS device is then fabricated by conventional means.

Figure 1–22 shows the chip layout for a completed monolithic, diode-protected, dual-gate MOSFET. In this structure, one of the diodes of each pair has been located under the gate bonding pads. The small

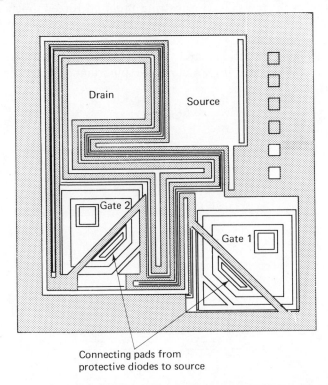

Connecting pads from
protective diodes to source

Fig. 1–22 Chip layout for a completed monolithic, diode-protected, dual-gate MOSFET. (Courtesy RCA)

triangular metal pads make contact with the second diode of each pair, and connect it to the source metalization. In assembly, the source is shorted to the substrate so that a low-resistance path to ground is provided for the diodes. To ground the diodes under the second gate properly, it is necessary to break the metal of the first gate, and terminate the first channel on the $P+$ guard band surrounding the diode structure of the second gate. This technique prevents spurious source-to-drain current which could result from the open nature of the structure.

The MOSFET depicted in Figs. 1–21 and 1–22 is a dual-gate device. Such a MOSFET can be converted to a single-gate device, but with diode gate protection, by a simple modification as shown in Fig. 1–23. This modification is accomplished by connecting the two gates [Fig. 1–23(a)]. This produces the electrical equivalent [Fig. 1–23(b)].

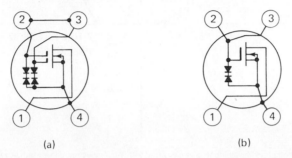

(a)   (b)

Fig. 1–23 Dual-gate MOSFET modified to provide single-gate operation with diode protection. (Courtesy RCA)

1–5.4. Characteristics of MOS Devices
Protected by Back-to-Back Diodes

No matter what fabrication technique is used, there will be some effect on the MOS device when protective diodes are added. However, the change in characteristics need not be significant. For example, with a typical RCA MOSFET operating as an RF amplifier at 200 MHz, the reduction in power gain and the increase in noise figure is less than 0.5 dB when gate protection diodes are added.

With a fixed drain current of 10 mA and a drain-to-source voltage of 15V, the input capacitance increased from about 5 pF to 8 pF when diodes were added, using the same MOSFET and operating frequency. Under the same test conditions, the input resistance decreased from about 1000 ohms to 800 ohms, with the addition of diodes. (Note that all test procedures for MOS devices are described in Chap. 6.)

The purpose of the protective diodes is to limit the amplitude of a transient to a value that is below the gate breakdown voltage, even in the presence of heavy input current surge. Figure 1–24 shows a typical diode transfer characteristic measured with a one-microsecond pulse at a 4×10^{-3} duty cycle. The MOSFET under test has a typical gate-to-source breakdown rating of 20V. The curves in Fig. 1–24 show that the transfer characteristic of the signal-limiting diodes will constrain a transient impulse to potential values well below the 20V limit, even when the input surge is capable of delivering hundreds of milliamperes.

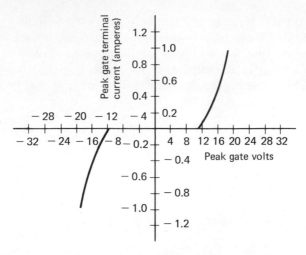

Fig. 1–24 Typical diode transfer characteristic measured with 1-microsecond pulse width at a 4×10^{-3} duty cycle. (Courtesy RCA)

1–5.5 Gate Protection for Complementary MOS Devices

The complementary MOS devices found in IC form also require protection. Generally this is done by fabricating a diode (or diodes) as part of the monolith chip. The protection scheme used by Motorola in their complementary MOS ICs (Sec. 1–3 and Chap. 5) is illustrated in Fig. 1–25.

Since the P-tub and N-substrate are lightly doped, the junction breakdown is high, about 120V. Therefore, a heavily-doped $N+$ region and

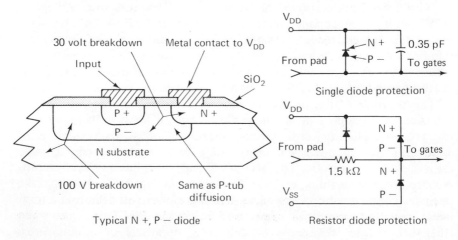

Fig. 1–25 Gate protection for complementary MOS devices. (Courtesy Motorola)

a lightly-doped *P*-region are used for the diodes. The junction between these two regions (*N*+ and *P*−) breaks down at about 30V which is well below the 100 to 120V gate-to-substrate breakdown.

The *single diode* scheme provides protection by clamping positive levels to V_{DD}. Negative protection is provided by the 30V reverse breakdown. The diode is designed to operate in the breakdown region without damage (provided the currents are kept under 10 mA).

The second method, known as *resistor diode protection*, adds some delay, but provides protection by clamping positive and negative potentials to V_{DD} and V_{SS} respectively. The resistor is included to provide additional circuit isolation.

1–6. CHARACTERISTICS OF MOS DEVICES

The various characteristics of MOS devices are discussed in detail throughout the other chapters of this book. The effect of these characteristics on certain applications of specific MOS devices is treated in detail. For example, drain-substrate junction capacitance is of great importance to a MOS device used in switching circuits, but is of no particular concern to a MOSFET operating as an RF amplifier.

In this section, we outline all the characteristics that generally apply to MOS devices of all types. This provides a foundation for understanding how the characteristics affect design with MOS devices. Both static and dynamic characteristics are discussed. Note that the various characteristics described here can be measured and tested. All test information, including test circuits and procedures, is included in Chap. 6.

1–6.1 Static MOS Characteristics

Static (or direct-current) characteristics of a MOSFET are those that indicate the effect of a control (or gate) voltage (or signal) on the output (drain-source) current. The following paragraphs describe static characteristics found on typical MOSFET data sheets. Figure 1–26 shows the relationship between the characteristics and typical $V_{GS} - I_D$ (gate-source voltage versus drain current) curves. Note that some of the characteristics must be related to the *operating mode* of the MOS device. For example, I_{DSS} is the drain current found with the MOSFET connected as a common-source device, but with *zero gate voltage*. As in Fig. 1–26, I_{DSS} for a MOS device in the depletion-only mode is essentially the maximum drain current (or the drain current with the device full-on). In the depletion-enhancement mode,

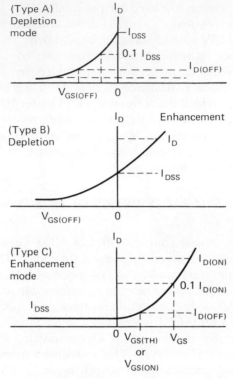

Typical $V_{GS} - I_D$ curves for
depletion mode RCA 3N128

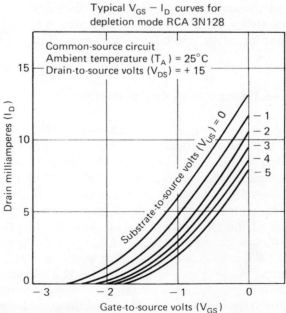

Fig. 1–26 Static MOS characteristics.

I_{DSS} is the drain current about half way between full-off and full-on. In the enhancement-only mode, I_{DSS} is essentially leakage current.

Control Voltage. The *gate-source voltage,* or V_{GS} is considered as the control voltage (or signal). That is, the amount of V_{GS} controls the amount of source-drain current.

$V_{GS(off)}$ is the gate voltage necessary to reduce the drain-source current (generally specified as I_D) to zero, or to some specified value near zero. Often, $V_{GS(off)}$ is defined as the gate voltage required to reduce the drain current to 0.01 or preferably 0.001 of the minimum I_{DSS} value. For example, assume that 1 mA flows when V_{GS} is at zero. This means that I_{DSS} is 1 mA. Under such conditions, $V_{GS(off)}$ is the gate voltage required to produce a drain current of 0.01 mA (or 0.001 mA).

Although the term $V_{GS(off)}$ can be applied to all operating modes, it is best applied to depletion and depletion-enhancement modes. In the enhancement-only mode, the term $V_{GS(th)}$ or *threshold gate voltage* is more descriptive. $V_{GS(th)}$ is the gate voltage of an enhancement-only mode MOS device where I_D just starts to flow. Some datasheets specify $V_{GS(off)}$ as the gate voltage that produces some given value of I_D, such as 1 μA, 1 pA, 1 nA, etc. The term $V_{GS(on)}$ is sometimes used in place of $V_{GS(th)}$ for enhancement-only mode devices.

No matter which term is used, the value of $V_{GS(off)}$ determines the *minimum operational* drain-source voltage, V_{DS}. As a rule of thumb, V_{DS} must be a minimum of 1.5 times $V_{GS(off)}$ [or $V_{GS(th)}$, or $V_{GS(on)}$].

Some datasheets specify $V_{GS(off)}$ as minimum-typical-maximum values, while other datasheets specify only a maximum $V_{GS(off)}$. There is a simple method of finding a workable value for $V_{GS(off)}$, if the $V_{GS} - I_D$ curves are available. Simply lay a straightedge along the straight portion of the curve as shown in Fig. 1–26. Note where the straightedge intersects the V_{GS} axis. Double this value to find a workable V_{GS}. For example, in the curve of Fig. 1–26, the straightedge intersects the V_{GS} axis at about -2V. Thus, a workable value for V_{GS} is -4V. Note that the datasheet for MOSFET with the Fig. 1–26 curve specifies $V_{GS(off)}$ as -0.5V minimum, -4V typical, and -8V maximum.

Operating Voltage. The *drain-source* voltage, or V_{DS} is considered as the operating voltage. V_{DS} is the equivalent of collector voltage in a two-junction transistor, or plate voltage in a vacuum tube. Generally, the only concern is that the maximum datasheet value for V_{DS} must not be exceeded. However, in most applications, V_{DS} must be a minimum of 1.5 times $V_{GS(off)}$.

When MOS devices are used as switches or choppers, the terms $V_{DS(on)}$ and $V_{DS(off)}$ sometimes appear on the datasheets. $V_{DS(on)}$ is similar to saturation voltage in conventional transistors. $V_{DS(off)}$ is the drain-source voltage where the I_D increases very little for an increase in drain-source voltage, with V_{GS} held at zero. Sometimes, the term V_P or pinch-off voltage is used instead. However, V_P generally applies to JFETs, rather than to MOS devices.

The various values of V_{DS} are dependent upon source resistance R_S, drain resistance R_D, or drain-to-source resistance R_{DS}. All of these are static, direct-current values, and rarely appear on datasheets. The dynamic r_{ds} values discussed in Sec. 1–6.2 are of far more importance.

Although MOS devices usually do not have an operating voltage between the drain and gate, the term V_{DG} generally appears on most MOS datasheets. This is a maximum voltage, and is usually the same as V_{DS} maximum.

Operating Current. The *drain-source* current, or I_D is considered as the operating current. I_D is the equivalent of collector current in a two-junction transistor, or plate current in a vacuum tube. Generally, the only concern is that the maximum datasheet value for I_D must not be exceeded.

In addition to I_D, the terms $I_{D(on)}$ and $I_{D(off)}$ are used on some datasheets. $I_{D(on)}$ is an arbitrary current value (usually near the maximum-rated current) that locates a point in the enhancement operating mode, as in Fig. 1–26. $I_{D(off)}$ is the current value that flows when $V_{GS(off)}$ is applied. On some MOS devices operating in the enhancement-only mode, the datasheet value of $I_{D(off)}$ is the current when $V_{GS(th)}$ is applied.

Voltage Breakdown. Voltage breakdown is a particularly important characteristic of MOS devices. There are a number of specifications to indicate the maximum voltage that can be applied to the various elements. These include:

$V_{(BR)GSS}$ —gate-to-source breakdown voltage
$V_{(BR)DGO}$ —drain-to-gate breakdown voltage
$V_{(BR)DSX}$ —drain-to-source breakdown voltage
$V_{(BR)DSS}$ —drain-to-source breakdown voltage (alternate specification)

$V_{(BR)GSS}$ is the breakdown voltage from gate to source, with the drain and source shorted. Under these test conditions, the gate-channel junction (oxide layer) also meets the breakdown specification since the drain and source are the connections to the channel. For the designer, this means that the drain and source may be interchanged, for symmetrical devices, without fear of individual junction breakdown. For MOS devices, the gate is insulated from the drain-source and channel by an oxide layer. Thus, the gate breakdown voltage is dependent on the thickness and purity of the layer. Gate breakdown voltage is the voltage that will physically puncture the oxide layer.

$V_{(BR)DGO}$ is essentially the same specification as $V_{BR(GSS)}$, except that $V_{(BR)DGO}$ represents breakdown from gate to drain. $V_{(BR)DGO}$ is more properly applied to JFETs, but may appear on a MOSFET datasheet.

Drain-to-source breakdown voltage for a MOS device is determined by the operating mode. For enhancement-only mode operation (Type *C*), with the gate connected to the source (in the cutoff condition) and the substrate floating, there is no effective channel between drain and source, and the applied drain-source voltage appears (in effect) across two back-to-back connected, reverse-biased diodes. These diodes are represented by the source-to-substrate and substrate-to-drain junctions.

During drain-to-source breakdown tests, drain current remains at a very low (typically picoamperes) level as drain voltage is increased until the drain voltage reaches a value that causes reverse (*reach-through* or *punch-through*) breakdown of the diodes. This condition, generally represented on datasheets as $V_{(BR)DSS}$, is indicated by an increase in I_D.

For types *A* and *B* MOS devices, the $V_{(BR)DSS}$ designation is sometimes replaced by $V_{(BR)DSX}$. The main difference between the two designations is the replacement of the last subscript "*S*" with the subscript "*X*". The *S* normally indicates that the gate is shorted to the source; the *X* that the gate is biased to cutoff, or beyond. To obtain cutoff in types *A* and *B* MOS devices, a depletion bias voltage must be applied to the gate.

In some datasheets, there may be ratings and specifications indicating the maximum voltages that may be applied between the individual gates and the drain or source, between the drain and source, and so on. Not all of these specifications are found on every datasheet. Some of the specifications provide the same information in slightly different form. However, by understanding the various breakdown mechanisms, it should be possible to interpret the intent of each specification and rating. For example, in MOSFET specifications the breakdown voltage is generally interchangeable with the term *avalanche* voltage (V_A). Avalanche voltage is the V_{DS} that will cause the I_D to go "full on", sometimes to the point of destroying the device. Note that avalanche occurs at a lower value of V_{DS} when the gate is reverse-biased, than when the gate is zero biased. This condition is caused by the fact that the reverse-bias gate-voltage adds to the drain voltage, thus increasing the effective voltage across the junction. As a rule-of-thumb, the maximum V_{DS} that may be applied is the drain-gate breakdown voltage, less the V_{GS}.

Gate Leakage. Gate leakage is a particularly important characteristic of a MOS device, since such leakage is directly related to input resistance. When gate leakage is high, input resistance is low, and vice versa. Gate leakage is usually specified as I_{GSS} (reverse-bias, gate-to-source current with drain shorted to source), and is a measure of the static short-circuit input impedance.

Normally, input resistance of a MOS device is very high since the leakage current across a capacitor is very small. At a temperature of

25°C, typical MOSFET input resistance R_{GS} is in the order of 10^{14} ohms. Input resistance may decrease when temperature is raised to the typical maximum of +175°C. However, MOS devices are not drastically affected by temperature, and their input resistance remains extremely high, even at maximum operating temperature.

Some datasheets specify gate leakage current as I_{GDO} (leakage between gate and drain with the source open). Others use the term I_{GSO} (leakage between gate and source with the drain open). When these characteristics are used, the gate leakage current is lower than when I_{GSS} is specified. Consequently, I_{GDO} and I_{GSO} does not represent the "worst case" condition, making I_{GSS} the preferred specification.

Device Dissipation. Datasheets list device dissipation (P_D) or transistor dissipation (P_T), or some similar term, at a given ambient temperature, together with a derating factor. Typically, P_D or P_T is the maximum power that can be dissipated with the device at 25°C without exceeding the maximum allowable internal temperature. Most MOS devices can not stand temperatures above about 200°C without damage. In some cases, the maximum temperature is 150 to 175°C.

A typical MOSFET will have a maximum temperature of +175°C, a maximum dissipation of 330 mW at 25°C, and a derating factor of 2.2 mW/°C. Thus, at 125°C, the power dissipated in the MOSFET must not exceed 110 mW (125°C − 25°C = 100°C; 100 × 2.2 mW = 220 mW; 330 mW − 220 mW = 110 mW).

Temperature Ranges. Critical temperatures for MOS devices are usually the *ambient temperature* during both operation and storage, as well as *lead temperature* during soldering. As a guideline, a typical ambient temperature range for MOS devices during both operation and storage is −65°C to +175°C. Lead temperatures should not exceed +265°C during soldering, and this temperature should not be closer than 1/32–inch from the seating surface for more than 10 seconds.

Protective Diode Voltage. On those MOS devices with protective diodes described in Sec. 1–5, the datasheets list the protective diode clamp voltage. This voltage is often referred to as the *knee* voltage, or V_{knee}. The input signal voltages can not exceed this value.

Forward Gate Current. Some MOS device datasheets list a forward gate current characteristic, such as I_{GS} or $I_{G(f)}$. This is not to be confused with I_{GSS} which is leakage current. I_{GS} or $I_{G(f)}$ is the maximum recommended forward current through the gate terminal. Although I_{GS} or $I_{G(f)}$ is primarily a JFET specification, the forward gate current characteristic can be a limiting factor in some MOS applications. I_{GS} or $I_{G(f)}$ is caused by a large forward bias current on the gate. When such a bias must be used, the gate current must be limited to the I_{GS} or $I_{G(f)}$ value (or degeneration of the device could occur). A resistor in series with the gate will limit current, but the resistor value will determine the variation of gate bias, as affected by the gate leakage current, and will affect input resistance (or impedance) of the device.

Dual Gate Characteristics. In addition to the static characteristics described thus far, dual-gate MOS devices have certain other characteristics. Both gates control I_D. As a result, there are a number of datasheet specifications that describe how the voltage on one gate will affect I_D, with the other gate held at some specific voltage, or zero volts. Likewise, there are specifications that describe how gate leakage or gate current flow is affected by gate voltages. For practical design situations, the most important dual-gate characteristics include cutoff voltage, breakdown voltage, and gate current.

With a dual-gate MOSFET, the gate 1 to source cutoff voltage (with gate 2 connected to the source) is identified as $V_{G1S(off)}$; the gate 2 to source cutoff voltage (with gate 1 connected to the source) is identified as $V_{G2S(off)}$. Note that dual-gate MOS devices are not always symmetrical. That is, the gate voltages in different combinations produce different drain currents. For example, with the RCA 40822 dual-gate MOSFET, $V_{G1S(off)}$ is tested with a V_{DS} of +15V, and a V_{G2S} of +4V. Under these conditions, a typical cutoff voltage of −2V at gate 1 will produce an I_D of 50 $\mu\Lambda$ [which is considered to be $I_{D(off)}$ for this depletion type device]. With the same MOSFET, $V_{G2S(off)}$ is tested with a V_{DS} of +15V, and a V_{G1S} of 0V. Under these conditions, a −2V at gate 2 will produce the I_D of 50 μA.

The gate 1 to source breakdown voltage (with gate 2 connected to the source) is identified as $V_{(BR)G1SS}$; gate 2 to source breakdown voltage is $V_{(BR)G2SS}$. The gate 1 leakage current is identified as I_{G1SS}; gate 2 leakage current is I_{G2SS}. With the RCA 40822, $V_{(BR)G1SS}$ is tested with a V_{DS} and a V_{G2S} of 0V. Under these conditions, a V_{G1S} of 9V will produce an I_{G2SS} of 100 μA.

With a dual-gate MOS device, gate leakage should be symmetrical, even though the gate-voltage versus drain-current characteristics may not be symmetrical. Thus, for the RCA 40822 with a V_{DS} and V_{G1S} of 0V, a V_{G2S} of 9V will also produce an I_{G2SS} of 100 μA. Further, gate leakage or gate current should be the same when gate voltages are reversed in polarity. Thus, dual-gate MOSFET datasheets often specify forward and reverse breakdown voltages as well as gate leakage currents.

As is the case with single-gate, dual-gate MOS devices have a maximum recommended gate current. This sets the limit of gate voltage. For example, the RCA 40822 has a maximum recommended gate current of 50 nA. A maximum of 50 nA gate current will flow when V_{DS} and one gate are at 0V, and the opposite gate is at 6V (either +6V or −6V). Thus, 6V is the maximum recommended gate voltage. Gate currents for dual-gate MOSFETs are usually identified as I_{G1SSF} and I_{G2SSF} (for forward) or I_{G1SSR} and $_{G2SSR}$ (for reverse).

Dual-gate MOS devices have a zero-bias drain current specification I_{DS} which corresponds to a I_{DSS} for single-gate units. However, the measurement for I_{DS} is usually based on 0V at one gate, and a fixed voltage at the opposite gate. For example, with the RCA 40822, I_{DS} is measured with V_{DS} at +15V, V_{G1S} at 0V, and V_{G2S} at +4V. Typically, the I_{DS} is between 5 and 30 mA for this depletion mode device.

1–6.2 Dynamic MOS Characteristics

Unlike the static characteristics, the dynamic characteristics (ac or signal) of MOS devices apply equally to Types *A*, *B* and *C*. However, conditions and presentations of the dynamic characteristics depend mostly on the intended *application*. Figure 1–27 indicates the dynamic characteristics generally used to describe a MOS device for various applications.

Audio	RF – IF	Switching	Chopper
Y_{fs} (1 kHz)	Y_{fs} (1 kHz)	C_{iss}	C_{iss}
C_{iss}	C_{iss}	C_{rss}	C_{rss}
C_{rss}	C_{rss}	$C_{D(sub)}$	$C_{D(sub)}$
Y_{os} (1 kHz)	G_{PS}	$R_{ds\,(on)}$	$R_{ds\,(on)}$
NF	MAG	t_d, t_s	
	MUG	t_r, t_f	
	Re (Y_{fs})(HF)		
	Re (Y_{is})(HF)		
	Re (Y_{os})(HF)		
	NF		

Fig. 1–27 Important dynamic characteristics generally used to describe a MOS device for various applications.

Forward Transadmittance (Transconductance). Forward transadmittance Y_{fs} is defined as the magnitude of the common-source forward transfer admittance. Y_{fs} is the most important dynamic characteristic for MOS-FETs, regardless of application. It serves as a basic design parameter in audio and RF, and is a widely accepted device figure of merit.

Because MOSFETs have many characteristics similar to those of vacuum tubes, the symbol g_m is sometimes used instead of Y_{fs}. This is further confused, since the g notation school also uses a number of subscripts. In addition to g_m, some datasheets show g_{fs}, while others go even further out with g_{21}.

No matter what symbol is used, Y_{fs} defines the relationship between an input signal voltage and an output signal current, with the drain-source voltage held constant, or:

$$Y_{fs} = \frac{\Delta I_D}{\Delta V_{GS}}, \qquad \text{with } V_{DS} \text{ held constant.}$$

Y_{fs} is expressed in mhos (current divided by voltage). In most data-sheets, Y_{fs} is specified at 1 kHz with a V_{DS} the same as that for which $I_{D(on)}$ or I_{DSS} is obtained. At 1 kHz, Y_{fs} is almost entirely real. Thus, Y_{fs} at 1 kHz equals Y_{fs}. At higher frequencies, Y_{fs} includes the effects of gate-to-drain capacitance and may be misleadingly high. For high frequency operation, the real part of transadmittance $R_e(Y_{fs})$, as discussed in later paragraphs should be used.

Since the $I_D - V_{GS}$ curves of a MOSFET are nonlinear, Y_{fs} will vary considerably with changes in I_D. This variation for a typical N-channel MOSFET (the 3N152) is illustrated in Fig. 1–28. Note that the curve ends sharply where Y_{fs} is at its maximum. This is the point where $I_D = I_{DSS}$.

Three Y_{fs} measurements are often specified for dual-gate MOS de-vices. One of these measurements, with two gates tied together, pro-vides a Y_{fs} value for the condition where a signal is applied to both gates simultaneously. The other two measurements provide the Y_{fs} for the two gates individually. Generally, with the two gates tied together, Y_{fs} is higher, and more gain may be realized in a given circuit. However, because of the increased capacitance, the gain-bandwidth product is much lower.

For MOSFETs used at radio frequencies, an additional value of Y_{fs} is often specified at or near the highest frequency of operation. This value is measured at the same voltage conditions as those used for $I_{D(on)}$ or I_{DSS}.

Because of the importance of the imaginary component at radio fre-quencies, the high-frequency Y_{fs} specification is generally a complex representation. That is, both forward transconductance (g_{fs}, g_{21}, etc.) and forward susceptance (b_{fs}, b_{21}, etc.) are shown, as in Fig. 1–29, which illustrates the curves for a 3N154 at a test frequency of 200 MHz.

Some datasheets list the real part of Y_{fs} (or forward transconduc-tance) as $Re(Y_{fs})$ or $Re(Y_{fs})(HF)$. Such $Re(Y_{fs})$ is defined as the com-mon source forward transfer conductance (drain current versus gate voltage). For high frequency applications, $Re(Y_{fs})$ is considered a sig-nificant figure of merit. The dc operating conditions are the same as for Y_{fs}, but the test frequency is typically 100 or 200 MHz. Rather than listing a value for the real part of Y_{fs} at a specific frequency, some datasheets show both the forward transconductance (g_{fs}) and the for-ward susceptance (b_{fs}) over the useful frequency range of the device. Such a representation is shown in Fig. 1–30 by means of curves. No matter what datasheet system is used, an increase in the real part of

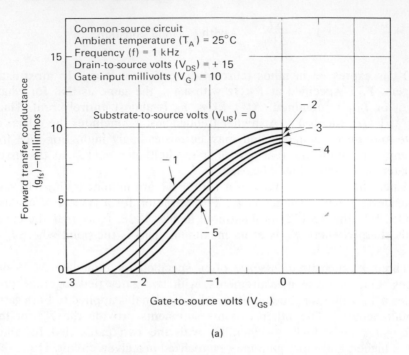

Common-source circuit
Ambient temperature $(T_A) = 25°C$
Frequency $(f) = 1\ kHz$
Drain-to-source volts $(V_{DS}) = +15$
Gate input millivolts $(V_G) = 10$

Substrate-to-source volts $(V_{US}) = 0$

Forward transfer conductance (g_{fs})—millimhos

Gate-to-source volts (V_{GS})

(a)

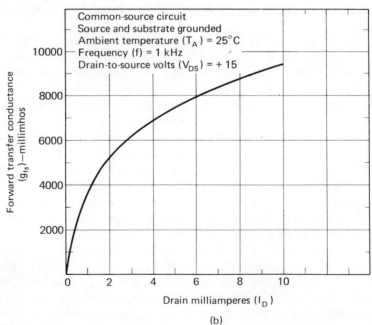

Common-source circuit
Source and substrate grounded
Ambient temperature $(T_A) = 25°C$
Frequency $(f) = 1\ kHz$
Drain-to-source volts $(V_{DS}) = +15$

Forward transfer conductance (g_{fs})—millimhos

Drain milliamperes (I_D)

(b)

Fig. 1–28 Forward transconductance versus (a) gate bias voltage and (b) drain current, for 3N152. (Courtesy RCA)

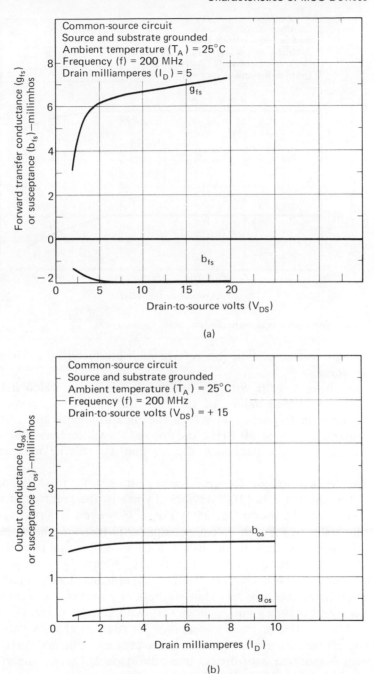

Fig. 1–29 Forward transadmittance versus (a) drain-to-source voltage and (b) drain current, for 3N154.(Courtesy RCA)

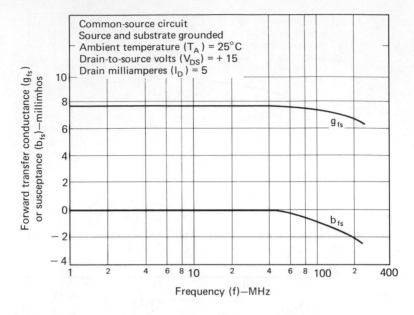

Fig. 1–30 Forward transadmittance versus frequency for 3N152. (Courtesy RCA)

Y_{fs} produces an increase in the voltage gain produced by a MOSFET amplifier stage.

In comparing $Re(Y_{fs})$ with Y_{fs} (on those datasheets which list both values) the minimum values of the two are quite close, considering the difference in frequency at which the measurements are made. At high frequencies, about 30 MHz and above, Y_{fs} will increase due to the effect of gate-drain capacitance C_{gd} so that Y_{fs} will be misleadingly high.

Reverse Transadmittance. Reverse transadmittance Y_{rs} (or Y_{12}) not generally a critical factor in MOS devices. Typically, the real part of Y_{rs} (or g_{rs}) is zero over the useful frequency range. However, it is necessary to know the values of Y_{rs} to calculate impedance-matching networks for MOSFET RF amplifiers, as discussed in Chap. 3. For that reason, most MOSFET datasheets list some values for Y_{rs}. Figure 1–31 shows some typical Y_{rs} curves. Note that the real part g_{rs} remains at zero for all conditions and at all frequencies. That is, there is no reverse conductance. However, the imaginary part b_{rs} does vary with voltage, current and frequency. That is, the reverse susceptance (or reactance) does vary with conditions. Thus, under the right conditions, there can be undesired feedback from output to input (reverse transadmittance). This condition must be accounted for in the design of RF amplifier stages.

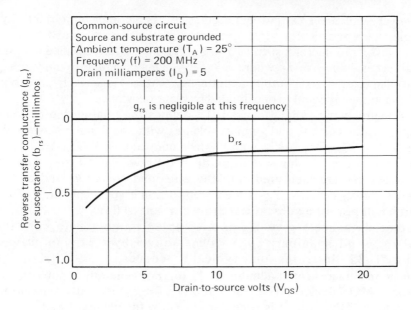

(a)

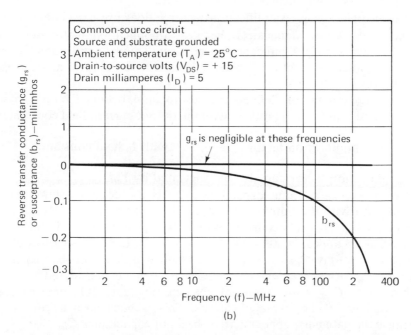

(b)

Fig. 1–31 Reverse transadmittance versus (a) drain-to-source voltage, and (b) frequency. (Courtesy RCA)

Output Admittance. Output admittance Y_{os} is another important dynamic characteristic for MOS devices. Y_{os} is also represented by various Y and g parameters, such as Y_{22}, g_{os}, g_{22}. Y_{os} is even specified in terms of drain resistance, or r_d, where $r_d = 1/Y_{os}$. This is similar to the vacuum tube characteristic of output admittance, where Y_{output} is $1/r_p$, or output admittance is equal to one divided by plate resistance.

No matter what symbol is used, Y_{os} defines the relationship between output signal current and output voltage, with the input voltage held constant, or $Y_{os} = \Delta I_D / \Delta V_{DS}$, with V_{GS} held constant.

Y_{os} is expressed in mhos (current divided by voltage). Types A and B MOS devices are measured with the gates and source grounded. For Type C MOS devices, Y_{os} is measured at some specified value of V_{GS} which will permit a substantial drain current to flow.

Some datasheets give Y_{os} as a complex number with both the real (g_{os}) and imaginary (b_{os}) values shown by means of curves. Figure 1–32 illustrates some typical Y_{os} curves.

Input Admittance. Input admittance Y_{is} (or Y_{11}) generally is not a critical factor in MOS devices. The real part of Y_{is} (or g_{is}) is nearly zero at low frequencies. However, it is necessary to know the values of Y_{is} to calculate impedance-matching networks for MOSFET RF amplifiers. Figure 1–33 shows some typical Y_{is} curves.

Amplification Factor. Amplification factor μ does not usually appear on most MOS device datasheets, because amplification factor does not usually have a great significance in most small-signal applications. However, amplification factor is sometimes used as a figure of merit in isolated cases.

Amplification factor defines the relationship between output signal voltage and input signal voltage, with the output current held constant, or

$$\text{Amplification factor} \quad = \frac{\Delta V_{DS}}{\Delta V_{GS}}, \quad \text{with } I_D \text{ held constant.}$$

Amplification factor can also be calculated by Y_{fs}/Y_{os}.

Input Capacitance. Input capacitance C_{iss} is the common-source input capacitance with the output shorted, and is used as a low-frequency substitute for Y_{is}. This is because Y_{is} is entirely capacitive at low frequencies.

To find an approximate value for Y_{is} at low frequencies (below about 1 MHz), multiply C_{iss} by 6.28F. The result will be b_{is}, or the imaginary part of Y_{is}. At these low frequencies g_{is} can be considered as zero.

Note that C_{iss} is an important characteristic for MOS devices used in switching or chopper applications. This is because a large voltage swing at the gate must appear across the input capacitance C_{iss}.

Output Capacitance. Output capacitance C_{oss} is the common-source output capacitance with the input shorted. C_{oss} does not appear on all MOS datasheets, but generally is found on dual-gate datasheets. The output

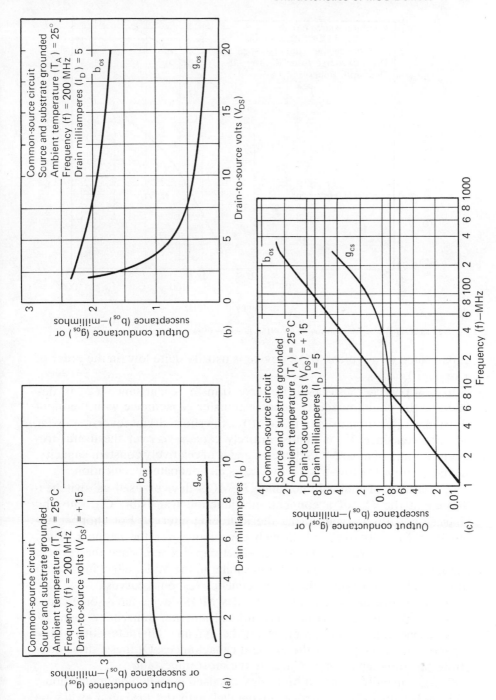

Fig. 1-32 Output admittance versus (a) drain current, (b) drain-to-source voltage, and (c) frequency for 3N152. (Courtesy RCA)

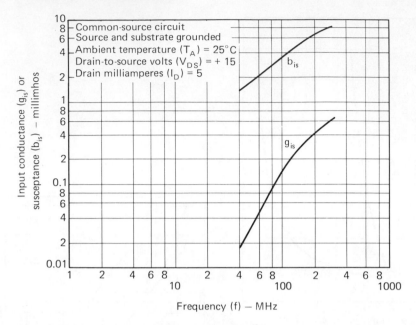

Fig. 1–33 Input admittance versus frequency for 3N152. (Courtesy RCA)

capacitance of a typical MOS device is usually quite low (in the order of 1 or 2 pF).

Reverse Transfer Capacitance. Reverse transfer capacitance C_{rss} is defined as the common-source reverse transfer capacitance with the input shorted. C_{rss} is often used in place of Y_{rs}, the short-circuit reverse transfer admittance, since Y_{rs} is almost entirely capacitive over the useful frequency range of most MOS devices, and is of relatively constant capacity. Consequently, the low-frequency C_{rss} is an adequate specification.

C_{rss} is also of major importance in MOS devices used as switches. Similar to the C_{ob} of a conventional junction transistor, C_{rss} must be charged and discharged during the switching interval. For chopper applications, C_{rss} is the feed-through capacitance for the chopper drive.

C_{rss} is also known as the Miller-effect capacitance, since the reverse capacitance can produce a condition similar to the Miller effect in vacuum tubes (such as constantly changing frequency response curves).

Element Capacitances. The elements of a MOS device have some capacitance between them, as do conventional junction transistors. Some of these capacitances have an effect on the dynamic characteristics of the device. Figure 1–6 shows the element capacitances. Of these, drain-substrate junction capacitance $C_{d(sub)}$ is the most important.

$C_{d(sub)}$ is usually found on MOS devices used in switching applications. This is because $C_{d(sub)}$ appears in parallel with the load in a switching circuit and must be charged and discharged between the two logic levels during the switching interval.

C_{ds} drain-to-source capacitance is another specification found on some switching MOS devices. C_{ds} also appears in parallel with the load in switching and logic applications.

Channel Resistance.   The channel resistance is an important characteristic for MOS devices used in switching applications. Channel resistance describes the bulk resistance of the channel *in series* with the drain and source. Channel resistance is described as $r_{d(on)}$, r_{DS}, R_{DS}, r_{ds}, $r_{d(off)}$, etc., depending on the datasheet. Some of these descriptions are static, some dynamic, and some mixed (again depending on the datasheet).

From a practical standpoint, there are two channel resistance specifications of concern in switching applications. These are the "on" and "off" specifications, such as $r_{ds(on)}$ and $r_{ds(off)}$. These can be either static or dynamic. The on specification is the channel resistance with the MOS device biased on. In a depletion mode device, the on condition can be produced by zero bias ($V_{GS} = 0$). In the enhancement mode, the on condition requires some forward bias. The opposite is true for the off condition. In a depletion device, the off condition requires some reverse bias. For enhancement, the off condition requires zero bias.

As a point of reference, a typical MOS device on resistance is in the order of 200 ohms, whereas the off resistance is greater than 10^{10}.

Switching Time.   Switching time characteristics are of particular importance for MOS devices used in digital applications (such as CMOS). Discrete MOSFETs are also used in switching applications. The datasheets that describe those MOSFETs specifically designed for switching include *timing characteristics*. Typically, these include t_d (delay time), t_s (storage time), t_r (rise time), and t_f (fall time). These characteristics are measured with a pulse source, and a multiple-trace oscilloscope, as are similar characteristics for conventional junction transistors.

Gain.   The datasheets for MOSFETs used as amplifiers usually list some gain characteristics. Typically, these are *power gain* figures, at some specified frequency and under specified test conditions (such as V_{DS}, I_D, etc.). The gain figures are expressed in dB, sometimes as maximum or minimum values. Such terms as MAG (maximum available gain), MUG (maximum usable gain), and G_{PS} (power gain with the device connected as common source) are used. The datasheet notes concerning gain should always be consulted. For example, the term MAG usually means the maximum available gain under ideal or theoretical conditions. The term MUG usually means the maximum usable gain when the amplifier is neutralized or mismatched to produce the stability effects of neutralization. On some datasheets, subscripts are added to identify the gain figures. For example, the subscript c can be added to MUG to indicate maximum usable gain when the MOSFET is used as a converter. (The problems of neutralization, mismatching and conversion are discussed in Chap. 3.)

Noise Figure. Noise figure for MOS devices is usually listed as NF on the datasheets, and represents a common-source noise figure. NF represents a ratio between input signal-to-noise ratio and output signal-to-noise ratio, and is measured in dB. In most MOSFET datasheets, NF includes the effects of e_n (equivalent short-circuit input noise expressed in volts per root cycle), and i_n (the equivalent open-circuit input noise current).

As shown in Fig. 1-34, NF attains its highest value for a small generator resistance, and decreases for increasing generator resistance, indicating a large noise contribution from the noise-voltage generator. For this reason, some datasheets specify both NF and e_n, but neglect i_n. In effect, NF is independent of operating current, and proportional to voltage. However, the voltage effects are slight over the normal operating range of a typical MOSFET.

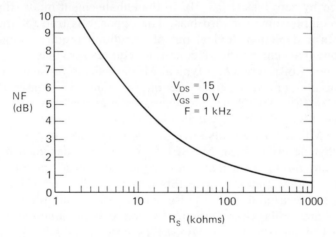

Fig. 1-34 Noise figure versus source resistance. (Courtesy Motorola)

Figure 1-35 is a nomograph for converting noise figure to equivalent input noise voltage for different generator source impedances R_S. This nomograph can be used with any MOSFET. Since NF and e_n are frequency dependent, Fig. 1-35 must be used in conjunction with Fig. 1-36 (noise figure versus frequency at a specified source impedance) to determine e_n. For example, to find the input noise at 50 Hz, with an R_S of 1 megohm, the NF (from Fig. 1-36) is about 1.5 dB. Next, from Fig. 1-35, noise voltage (e_n) is found to be 1.5×10^{-7} (on the 10^6 or one megohm curve) volts, or about 150 nV. If the stage has a voltage gain of 10, the output noise is 10×150 nV or 1.5 μV of 50 Hz noise. As another example, the NF for an RCA 3N154 is given as 5 dB (maximum) at 200 MHz. With the same one meghom source resistance R_S the noise voltage is approximately 2×10^{-7} (200 nV).

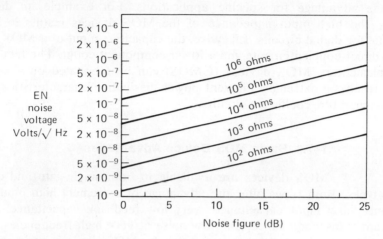

Fig. 1-35 Noise figure conversion chart. (Courtesy Motorola)

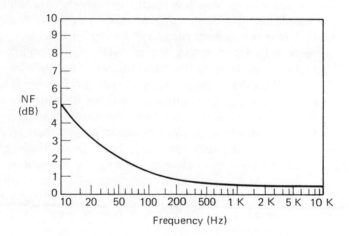

Fig. 1-36 Noise figure versus frequency. (Courtesy Motorola)

1-7. ADVANTAGES AND LIMITATIONS OF MOS DEVICES

From a study of the material presented thus far, it is obvious that MOS devices have many advantages over conventional junction transistors and JFETs. These advantages and limitations are discussed in detail throughout the remaining chapters. In this section, we shall summarize the advantages and limitations as they apply to typical applications.

As discussed later, many of the typical MOS characteristics provide definite advantage for specific applications. For example, in digital work, the high input impedance of the MOS devices results in high fan-out for digital circuits. Likewise, the capacitive input of a MOS permits direct-coupled circuits and a lower component count. The fact that complementary MOS devices (CMOS) can be fabricated on a single chip results in extremely efficient power dissipation characteristics and low temperature coefficient circuits.

1–7.1 Basic MOS Device Advantages

MOS devices are available in both single-gate and dual-gate types. Both types offer the advantage of extremely high input resistance, low input capacitance, very low feedback capacitance, high forward transconductance, and low noise at very high frequencies. Because of their insulated-gate construction, MOS devices have an extremely low leakage current which is relatively insensitive to temperature variations. In addition, the drain currents have a negative temperature coefficient which makes "thermal runaway" almost impossible.

The extremely high input resistance of MOSFETs permits the use of simple, electron-tube-type biasing techniques. The high input resistance also makes MOSFETs capable of handling relatively large positive and negative input-signal excursions without degradation of input impedance due to diode-current loading. Because of this capability, MOSFETs have considerably greater dynamic ranges than JFETs and conventional bipolar transistors. In turn, because of the higher dynamic range, MOSFETs can provide superior performance in amplifier circuits with AGC (automatic gain control). For example, because of the high input resistances, MOSFETs impose virtually no loading on AGC voltage sources (whereas a conventional junction transistor requires considerable AGC current).

Probably the greatest single advantage of MOS devices in digital and IC applications is the ability to fabricate a large number of devices on a single chip. For example, Texas Instruments has put up to 5000 devices on a silicon chip only 150×150 mils square.

In addition to these basic advantages, MOSFETs are notably superior to other solid-state devices in cross-modulation characteristics, and in their relative freedom from spurious responses. MOSFETs are also characterized by zero offset voltage; a feature making them especially desirable for chopper applications.

1–7.2 Typical Single-Gate MOSFET Ranges

Low Feedback Capacitance. All MOSFETs have low reverse or feedback capacitance C_{rss}. Single-gate MOSFETs have C_{rss} values in the range of 0.1 to 0.3 pF. This low feedback capacitance often eliminates the need for neutralization, or mismatching, in RF amplifiers. (As discussed in Chap. 3, conventional RF amplifiers often require neutralization or mismatching to prevent oscillation resulting from feedback between output and input). A high feedback capacitance can also result in Miller effect (constantly changing frequency response). A typical low-power bipolar transistor will have a reverse capacitance (generally listed as C_{ob}) of 5 to 10 pF, and higher.

The low C_{rss} of a MOSFET is also an advantage in switching applications, since C_{rss} must be charged and discharged on each cycle of switching. The lower the C_{rss} value, the higher the possible switching frequency.

Low Gate Leakage Current. Because of the insulated gate construction, all MOSFETs have low gate leakage current I_{GSS}. Single-gate MOSFETs have I_{GSS} values in the range of 1 nA, or less (often as low as 0.1 pA). Even at maximum operating temperatures I_{GSS} remains in the order of pA, rather than nA. In turn, low gate leakage currents result in high static input resistances or impedances, even at maximum operating temperatures.

The low gate leakage current also eliminates the problem of diode-current loading of the input circuit under strong input conditions, which is common to bipolar transistors and JFETs. With these other devices, a strong input signal will result in considerable input leakage current, thus loading the input signal source. MOSFET gate leakage remains low, even in the presence of maximum permitted input signal.

Wide Dynamic Range. Very little gate current is drawn by a MOS device in the presence of an input signal (either reverse or forward bias input). Since there is very little current loading, a MOS device will handle both positive and negative input-signal excursions over a wide range. A typical MOSFET will handle ac inputs of ±15V, and continuous dc inputs of ±10V.

Wide-range, low-power AGC—The wide input signal range also applies to AGC voltages, thus producing a wide AGC range. This is particularly useful in high-performance receiver applications. Because little gate current is drawn in either the forward or reverse input condition, MOSFET AGC systems are voltage-operated, rather than current-operated (as are most conventional junction transistor AGC systems). That is, MOSFET AGC systems operate on a feedback of voltage changes, rather than current changes. Thus, little power is required to operate MOSFET AGC systems.

High Forward Transconductance and Power Gain. MOS devices have a high forward transconductance, typically in the order of 7500μmhos (with 20,00 μmhos possible). Of course, there is no comparable specification for conventional junction transistors. However, forward transconductance (output current versus input voltage) is a measure of power gain. MOSFETs have power gains in the order of 15 to 20 dB.

Negative Temperature Coefficient. The drain current of a MOS device generally operates on a negative temperature coefficient (NTC). That is, as temperature increases, drain current decreases. This prevents thermal runaway common in junction transistors. With conventional transistors, an increase in temperature produces an increase in current. In turn, the current increase produces a further temperature increase. Unless precautions are taken, conventional junction transistors can be damaged or destroyed by thermal runaway. No such precautions are required with MOSFETs. It is only necessary to provide a sufficient heat sink to dissipate the maximum-rated power of the MOSFET. Generally, this is done quite simply by keeping the MOSFET case in good contact with the metal chassis.

Zero Temperature Coefficient. An important advantage of all FETs is their ability to operate at a zero-temperature-coefficient ($0TC$) point. This means that if the gate-source is biased at a specific voltage, and is held constant, the drain current will not vary with changes in temperature. Typically, JFETs show the $0TC$ characteristic over a wide range of temperature, approximately 150°C. MOSFETs are limited to a much narrower range, approximately 50°C. Likewise, it is not always practical to operate a MOSFET at some fixed V_{GS}. For these reasons, $0TC$ is not always considered an important MOSFET characteristic, even though it can be an advantage in certain applications. The $0TC$ characteristic is discussed thoroughly in Chap. 2.

Zero Offset Voltage. There is no inherent offset voltage associated with MOSFETs, as there is with conventional junction transistors. This is a particular advantage in chopper applications, or in amplifiers where MOS devices are used as the input stage. In a typical circuit, input offset voltage can be defined as that voltage that must be applied at the input terminals to obtain zero output voltage. With a MOSFET in a typical circuit, zero volts in produces zero volts out.

1–7.3 Dual-Gate MOS Device Advantages

A dual-gate MOS device is essentially a *series* arrangement of two separate channels, each channel having an independent control gate. This arrangement results in substantially lower feedback capacitance, greater gain, remote AGC capability in RF amplifier applications, substantially better cross-modulation characteristics, and lower spurious response than are provided by single-gate types. The availability of two independent control gates also offers unique advantages for chopper, clipper, and gated-amplifier service, and for applications involving the combination of two or more signals, such as mixers, product detectors, color demodulators, and balanced modulators.

1–7.4 Typical Dual-Gate MOSFET Ranges

Extremely Low Feedback Capacitance. The C_{rss} of a typical dual-gate MOSFET without diode protection is in the order of 0.02 to 0.03 pF. When protective diodes are added, the C_{rss} increases to the 0.03 to 0.05 range. However, this is still below the typical 0.1 to 0.3 pF range of a single-gate MOSFET, and well below the 5 to 10 pF C_{ob} of a junction transistor. The extremely low C_{rss} of dual-gate MOSFETs permit design of many RF amplifier circuits without neutralization, or mismatching, to produce stability.

The very low C_{rss} of dual-gate MOSFETs also permits operation at higher frequencies (to approximately 500 MHz), and prevents such conditions as oscillator feedthrough in converter circuits.

Remote AGC. Since a voltage applied to either gate affects the drain current, it is possible to incorporate remote AGC with dual-gate MOSFETs. That is, the signal can be applied to one gate, and the AGC feedback voltage applied to the other gate. This results in increased gain reduction for a given AGC voltage.

Exceptionally High Forward Transconductance. The forward transconductance of a typical dual-gate MOSFET is in the order of 10,000 to 15,000 μmhos, with 20,000 μmhos possible. This compares to about 7500 μmhos for single-gate MOSFETs.

Increased Power Gain. The power gain of a typical dual-gate MOSFET is in the order of 18 to 20 dB, with 28 to 30 dB possible. This compares to about 15 to 20 dB for single-gate MOSFETs.

1–7.5 MOS Device Limitations

As is the case with any electronic component, all MOS devices have certain characteristic limitations. In some applications these limitations present no particular problem. In different applications, the characteristics of MOS devices must be compared with characteristics of other devices (JFETs and conventional junction transistors) to make an intelligent selection. Of course, future design of MOS devices may be improved, thus removing the limitations. Until that time, the informed MOS user should consider the following:

Power Output. The maximum power dissipation of a typical MOS-FET is in the range of 300 mW, with 400 mW generally the top limit. Drain current (I_D) is typically 50 mA or less. Thus the maximum output power of present day MOSFETs is a fraction of one watt.

Operating Frequency. The maximum operating frequency of a typical MOSFET is in the range of 200 MHz, with 500 MHz generally the top. This compares to 1200 MHz (and higher) for conventional junction transistors.

Handling MOS Devices. As discussed in Sec. 1–4, MOS devices do require special handling when out of circuit. This means that extra care must be used during service, as well as assembly, of equipment with MOS devices. Special problems are created with automatic assembly equipment. All of the problems are considered as limitations by some technicians and engineers.

Durability. In circuit, MOS devices are about as rugged as JFETs and junction transistors, with one possible exception. When a MOSFET is subjected to an input voltage that causes the metal oxide gate to break down, the MOS device is destroyed. Other junction-type transistors can often withstand momentary overloads (voltage and/or current) without permanent damage.

Protective Diode Limitations. As discussed in Sec. 1–5, the problem of damage due to momentary input overloads can be eliminated by means of protective diodes. However, the addition of protective diodes presents an operating limitation on the MOS device. When the protective diodes are used, the maximum input signal (and dynamic range) is set by the diode, rather than the MOS characteristics. Typically, the protective diodes restrict the input to a range of about 10 or 15V.

The use of protective diodes also increases gate leakage current. The I_{GSS} of a typical MOS device without protective diodes is 1 nA (or less). With protective diodes, the I_{GSS} is increased to 50 nA.

Chopper Limitations. Both MOSFETs and JFETs have a low drain-source resistance when switched on. However, JFETs that are specifically designed for switch or chopper use have a lower $r_{ds(on)}$ than comparable MOSFETs. Typically, a chopper MOSFET will have an $r_{ds(on)}$ of about 200 ohms, whereas a JFET for chopper use will have an $r_{ds(on)}$ of about 25 ohms. Of course, this advantage is generally offset by the much higher capacitances and input leakage common to JFETs.

2. WORKING WITH DISCRETE MOS DEVICES

This chapter is devoted to the discrete MOS devices in common use. Specifically, it covers MOSFETs, both single-gate and dual-gate. Also discussed is the *triode-connected* dual-gate MOSFET. In addition to discussing the characteristics of discrete MOS devices, the chapter analyzes the various applications for which MOSFETs are best suited. Practical versions of the many MOSFET applications are discussed in Chap. 3 and 4.

2–1. BASIC MOSFET OPERATING REGIONS

Both MOSFETs and JFETs have three distinct characteristic regions, only two of which are operational. Figure 2–1(a), the output transfer characteristic of a depletion-type MOSFET or JFET, illustrates the three different regions. Below the pinch-off voltage V_P, the FET operates in the *ohmic* or *resistance region*. Above this voltage, up to the drain-source breakdown voltage $V_{(BR)DSS}$, the FET operates in the *constant current region*. Above the breakdown voltage, is the *avalanche region* where the FET is not operated. For this reason, the avalanche region is referred to by some manufacturers as the *forbidden region*.

The drain-source resistance r_{ds} at any point on these curves is given by the slope of the curve at that point. Above the pinch-off voltage, changes in the drain-source voltage V_{DS} result in small changes in drain current I_D. This produces a very high drain-source resistance and is characteristic of a constant-current source. Also, the actual operating drain current is variable and is dependent upon the gate-source voltage, resulting in a voltage-controlled current source.

The $I_D - V_{GS}$ curve, shown in Fig. 2–1(b), illustrates how the drain current varies with changes in gate-source voltage. For depletion FETs, the drain current decreases as the gate-source voltage is increased. For enhancement devices, the drain current is enhanced or increased as the gate-source voltage is increased.

If the FET is operated with a drain-source voltage below the pinch-off voltage, or preferably a voltage below several hundred millivolts, the slopes of the curves vary considerably as the gate-source voltage is changed. This is shown in Fig. 2–1(c). Since the slope varies, the drain-source resistance varies. This operation is in the *ohmic region*, where the drain-source channel is actually a voltage-variable or voltage-controlled resistor. As shown in Fig. 2–1(d), the drain-source resistance decreases with increasing gate-source voltage, for enhancement FETs. The converse is true for depletion FETs, MOSFETs are generally not operated in the ohmic region. However, MOSFETs are capable of operation below the pinch-off voltage.

Note that the curves in the ohmic region [Fig. 2–1(c)] are relatively symmetrical. This means that ac as well as dc signals can be handled. In other words, the drain-source channel is bilateral, not unilateral.

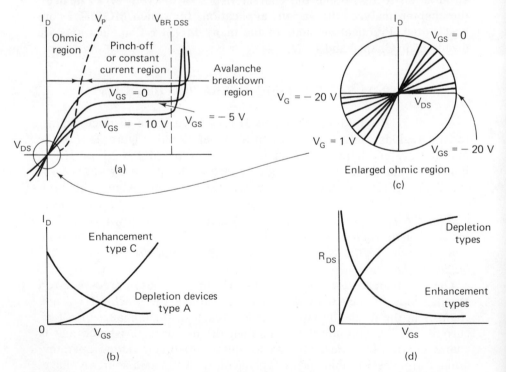

Fig. 2–1 Characteristics of FET operating regions. (Courtesy Motorola)

Also note that FETs are generally operated above the pinch-off region (in the constant-current region) for linear applications, whereas the ohmic region is used only for voltage-variable resistor applications.

Figure 2–2 illustrates I_D as a function of V_{DS} for a typical N-channel depletion-type MOSFET. Three I_D curves are shown; $V_{GS} = 0$, V_{GS} greater than 0, and V_{GS} less than 0. Changes in the conductivity pattern are shown in the simplified conductivity profile for each region of operation.

In the ohmic region, the $I_D - V_{DS}$ curve has the characteristics of a resistance. The shape of this curve is a function of V_{DS}. That is, the slope of the curve is governed by V_{GS}. The V_{DS}/I_D characteristic, or resistance, is controlled by the gate voltage.

As V_{DS} is increased, it produces an electrostatic stress in the channel. In turn, the stress modifies channel conductivity as shown. The channel is completely pinched off beyond V_P. Increasing V_{DS} serves only to maintain I_D at a constant level.

In the constant current (or *amplifier*) region, the I_D is held at a constant level for a given fixed-gate voltage, V_{GS}. A change in V_{GS} produces a change in I_D. Thus, in the constant current region, the device exhibits the transconductance characteristic that is essential to amplifier operation (that is, G_m or transconductance $= dI_D/dV_{GS}$).

In the forbidden region, an increase of V_{DS} beyond the rated maximum could produce avalanching in the drain-to-substrate diffusion (diode).

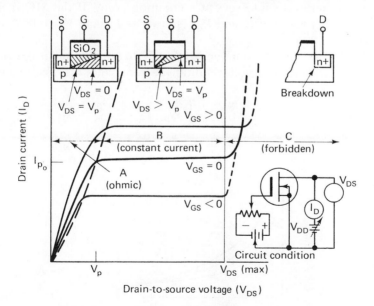

Fig. 2–2 Regions of operation for N–channel depletion MOSFET. (Courtesy RCA)

For this reason, MOSFETs of any type should not be operated in the forbidden region (above the maximum rated V_{DS}).

2–2. ZERO–TEMPERATURE–COEFFICIENT POINT

An interesting characteristic of all FETs is their ability to operate at a zero-temperature-coefficient ($0TC$) point. This means that if the gate-source is biased at a specific voltage, and is held constant, the drain current will not vary with changes in temperature.

Typically, JFETs show the $0TC$ characteristic over a wide range of temperature, approximately 150°C. MOSFETs are limited to a much narrower range, approximately 50°C. For that reason, $0TC$ is not generally considered as an important MOSFET characteristic.

Also, it is not always practical to operate a MOSFET at the $0TC$ point. For example, assume that the required V_{GS} to produce $0TC$ is 0.3V, and the MOSFET is to operate as an amplifier with 0.5V input signals. A part of the input signal will be clipped. Or, assume that the circuit is to be self-biased with a source resistor. An increase in bias resistance to produce $0TC$ could reduce gain. Thus, in practical circuits, MOSFETs are generally not biased for $0TC$. However, the $0TC$ characteristic should be understood and is worth discussing.

The $I_D - V_{GS}$ curves of Fig. 2–3 show that the various curves of different temperatures intersect at a common point. If the MOSFET is operated at this value of I_D and V_{GS} (shown as I_{DZ} and V_{GSZ}), $0TC$ operation will result.

The $0TC$ point varies from one MOSFET to another and is dependent upon I_{DSS}, the zero-gate-voltage drain current, and V_p. The equations

$$I_{DZ} \approx I_{DSS} \left(\frac{0.63}{V_p} \right)^2 \approx \frac{0.4\, I_{DSS}}{V_p^2}$$

$$V_{GSZ} \approx V_p - 0.63$$

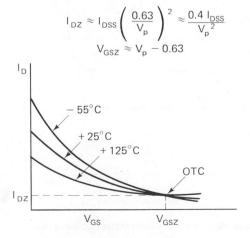

Fig. 2–3 Zero temperature coefficient for FETs. (Courtesy Motorola)

shown in Fig. 2–3 provide good approximations of the $0TC$ point. For example, if the pinch-off voltage V_p is 1V, the $0TC$ mode will be obtained if the gate-source voltage V_{GS} is 0.37V ($1 - 6.3 = 3.7$).

2–2.1 Practical Method for Finding 0TC

The values of I_D and V_{GS} that produce $0TC$ can be found by using datasheet curves, or by equations, as shown in Fig. 2–3; however, the values are typical approximations.

A more practical method for determining I_{DZ} requires a soldering tool, a coolant (a can of Freon), and a curve tracer (such as the Tektronix Type 575). By placing a 1000-ohm resistor across the base and emitter terminals of the curve tracer test socket, the constant-current base drive is converted to a relatively constant voltage for driving the MOSFET gate. The curve tracer is adjusted to display the I_D–V_{GS} output family [such as shown in Fig. 2–1(a)]. By alternately bringing the soldering tool near the MOSFET and spraying the MOSFET with Freon, the *voltage step of V_{GS} which remains motionless* on the curve tracer can be observed. The I_D at this voltage step is I_{DZ}.

Typically, FETs with an I_{DSS} of about 10 to 20 mA will have an I_{DZ} of less than 1 mA. Usually, the I_{DZ} increases as I_{DSS} increases (but not always, and not in proportion). For example, the I_{DZ} of a 50 mA MOSFET often shows an I_{DZ} below 1 mA.

2–3. TRIODE-CONNECTED DUAL-GATE MOSFET

It is possible to connect a dual-gate MOSFET so that it functions as a single-gate device, as discussed in Sec. 1–5.3. This is known as the *triode-connected* configuration and is shown in Fig. 2–4 which illustrates a dual-gate MOSFET with back-to-back diode protection converted to a "triode" with gate protection.

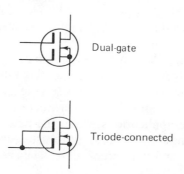

Dual-gate

Triode-connected

Fig. 2–4 Triode-connected dual-gate MOSFET.

The triode-connected configuration has curve tracer (drain family) characteristics that look like the "real" triode. The curves in Fig. 2–5 show that characteristics for the triode MOSFET (3N128) and the triode-connected dual-gate MOSFET (3N187) are essentially similar.

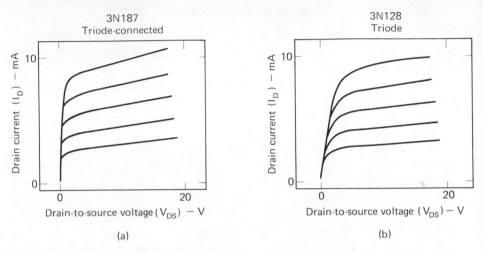

Fig. 2–5 Comparison of $V_{DS} - I_d$ curves for triode MOSFET and triode-connected MOSFET. (Courtesy RCA)

Some useful triode-connected device characteristics are provided in Fig. 2–6 in the form of comparisons with dual-gate and single-gate devices. It should be noted that the difference in I_{DS} level between the 3N187 and 3N200 carries over to their triode-connected versions. A curve showing I_{DSS} for triode-connection versus I_{DS} for the dual-gate configuration (with V_{G2S} at 4V) is shown in Fig. 2–7.

A plot of the triode-connected dual-gate transfer characteristics (I_D versus V_{GS}) is shown in Fig. 2–8. Similarly, g_{fs} curves are given in Fig. 2–9 as functions of I_D. Curves for typical dual-gate operation are available in commercial datasheets.

Typical variations in $R_{D(ON)}$ as a function of gate voltage are shown in Fig. 2–10. Note that $R_{D(ON)}$ is given (in ohms) for both tetrode-connected (separate gates) and triode-connected configurations. Both configurations compare favorably with single-gate devices.

It should not be inferred from these comparisons that all single-gate applications can be handled by protected (back-to-back diodes) dual-gate devices. As discussed in Sec. 1–7.5, gate leakage is increased when back-to-back diodes (or any form of protective diode) is added to a MOSFET. For those applications where the gate leakage I_{GSS} must be in the picoampere range, it is necessary to use a classic single-gate type (such as the 3N128), and take precautions against gate-insulation puncture.

Characteristic	Conditions	Triode-connected		Dual-gate circuit*		Single gate	Units
		3N187	3N200	3N187	3N200	3N128	
I_{DS}	$V_{DS} = 15$ V	6.0	2.0	15	5	15	mA
g_{fs}	$\begin{cases} V_{DS} = 15 \text{ V} \\ I_D = 10 \text{ mA} \\ f = 1 \text{ kHz} \end{cases}$	7.0	8.5	12	15	9	mmho
V_{G1S} (OFF)	$\begin{cases} V_{DS} = 15 \text{ V} \\ I_D = 50 \ \mu A \end{cases}$	−2.0	−1.0	−2.0	−1.0	− 1.5	V
I_{G1SS}	$V_{GS} = \pm 6$ V	2.0	2.0	1.0	1.0	10^{-4}	nA
C_{iss}	$\begin{cases} V_{DS} = 15 \text{ V} \\ I_D = 10 \text{ mA} \\ f = 1 \text{ kHz} \end{cases}$	10.0	10.0	6.0	6.0	5.5	pF
C_{rss}	$\begin{cases} V_{DS} = 15 \text{ V} \\ I_D = 10 \text{ mA} \\ f = 1 \text{ kHz} \end{cases}$	0.5	0.5	0.02	0.02	0.2	pF
C_{oss}	$\begin{cases} V_{DS} = 15 \text{ V} \\ I_D = 10 \text{ mA} \\ f = 1 \text{ kHz} \end{cases}$	2.0	2.0	2.0	2.0	1.4	pF
R_{DS} (ON)	$\begin{cases} V_{DS} = 1 \text{ V} \\ V_{GS} = 0 \end{cases}$	160	250	100	150	300	ohm

* $V_{G2S} = 4$ V except for I_{GSS} measurement, where $V_{G2S} = 0$

Fig. 2-6 Comparison of typical electrical characteristics for triode-connected dual-gate, standard dual-gate, and triode (single-gate) MOSFETS. (Courtesy RCA)

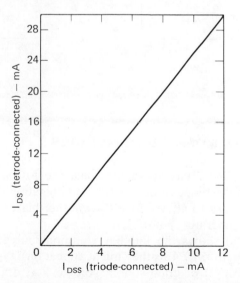

Fig. 2-7 Correlation of zero-bias drain current for the protected dual-gate device in tetrode (I_{DS}) and triode-connected (I_{DSS}) configurations. (Courtesy RCA)

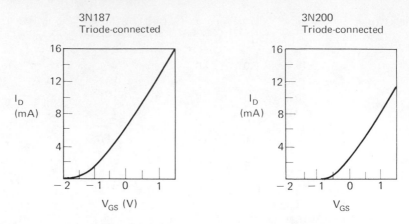

Fig. 2–8 Triode-connected dual-gate MOSFET transfer characteristics. (Courtesy RCA)

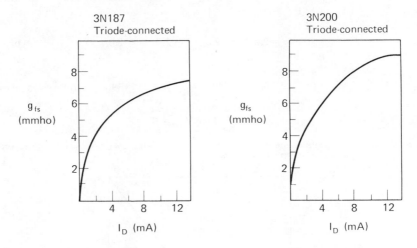

Fig. 2–9 Triode-connected dual-gate MOSFET transconductance characteristics. (Courtesy RCA)

2–4. MOSFET BIAS METHODS

In linear circuit applications, the MOSFET is biased by an external supply, by self-bias, or by a combination of these two techniques. This applies to all MOSFETs, whether biased at the $0TC$ point or at some other operating point.

Figure 2–11(a) shows the familiar common-source, drain characteristic curves for a MOSFET as they might appear on a typical curve tracer. For a constant level of drain-source voltage V_{DS}, drain current I_D can be plotted versus gate-source voltage V_{GS}, as shown in Fig. 2–11(b). This latter curve is generally referred to as the transfer characteristics.

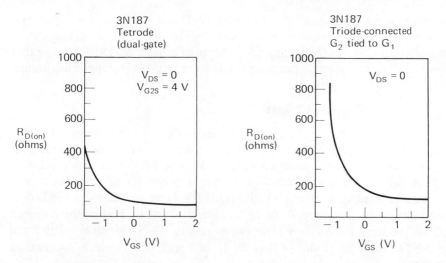

Fig. 2–10 "ON" resistance as a function of gate voltage for tetrode and triode-connected dual-gate MOSFET. (Courtesy RCA)

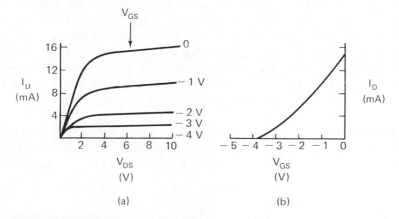

Fig. 2–11 MOSFET characteristic curves: (a) common-source drain characteristics, (b) transfer characteristics.

However, from a practical standpoint, either curve shows the amount of current that flows through the MOSFET for a given gate-source voltage. For example, either curve shows that if a −1V bias is applied between gate and source, approximately 10 mA will flow. If the supply voltage (drain-source voltage) is 10V, and a 500–ohm resistor is connected between drain and supply, there is a 5V drop across the resistor. Of course, this reduces the drain-source voltage down to 5V and possibly changes the characteristics. (In the example of Fig. 2–11(a), there would be very little change in characteristics.)

The following paragraphs provide methods by which similar curves can be used to find the correct value of bias for a given quiescent (no-signal) operating point of a MOSFET. These methods are basic and do not allow for variations in temperature. In Sec. 2–6, we discuss step-by-step procedures for finding bias values of a basic MOSFET stage that will maintain a given operating point over a given temperature range.

2–4.1 Self Bias

Figure 2–12 shows the basic circuit for self-bias. The design of a self-bias circuit is relatively simple. Generally, straight self-bias is used when the main concern is that the MOSFET will produce a given forward transconductance g_{fs} for a given set of conditions. For example, assume that a 3N128 MOSFET is to produce 7.4 mmhos at an ambient temperature T_A of 25°C. Design starts by finding the required I_D to produce 7.4 mmhos. The curve of Fig. 2–13, which is taken from a 3N128 datasheet, shows that an I_D of 5 mA will produce 7.4 mmhos when the V_{DS} is 15V and the T_A is 25°C. Note that V_{DS} must be 15V; thus, the supply or V_{DD} must be higher since there will be some drop across R_S. Also note that there is no resistance in the drain circuit.

In a practical circuit, there will always be some resistance in the drain to act as a load. An exception is the common-drain (source follower) circuit. When a drain resistance is used, the drop across the drain resistance must be considered to determine the supply voltage. All these factors are discussed in Sec. 2–6. However, for the purposes of this section, we are concerned only with the basic or theoretical calculations for the bias circuit.

The next step is to determine the required gate voltage V_{GS} to produce the I_D of 5 mA. The curve of Fig. 2–14, also taken from the 3N128 datasheet, shows that a V_{GS} of −1.1 will produce an I_D of 5 mA, provided

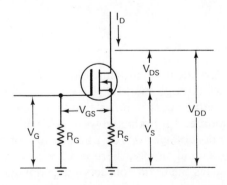

Fig. 2–12 Basic self-bias circuit.

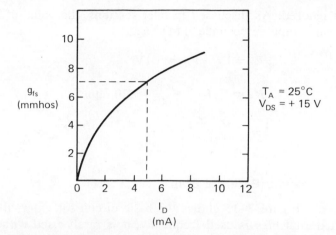

Fig. 2-13 Transfer characteristics for 3N128. (Courtesy RCA)

that the T_A remains at 25°C and that the device is "average". The -1.1V figure is taken from the solid curve of Fig. 2–14; the dashed curves show the possible high and low limits of I_D for a given V_{GS}.

Assuming that the solid curve is good and that the -1.1V figure will hold, the values of R_S and V_{DD} can be calculated as follows. Note that since no fixed bias is applied, V_G is considered as 0V. Also, the value

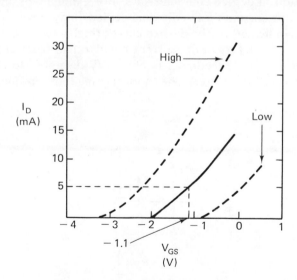

Fig. 2–14 Operating characteristics for 3N128. (Courtesy RCA)

of R_G is ignored. As discussed in later sections, the value of R_G is set by the desired input impedance of the stage.

$$V_S = V_G - V_{GS} = 1.1\text{V}$$

$$R_S = V_S/I_D = \frac{1.1}{0.005} = 220 \text{ ohms}$$

$$V_{DD} = V_{DS} + V_S = 15 + 1.1 = 16.1\text{V}$$

2–4.2 External Bias

Figure 2–15 shows the basic circuit for external bias. Although external bias is used frequently, it is rarely used without some form of self-bias. Thus, the circuit of Fig. 2–15 is theoretical. As in the case of self-bias, design starts by finding the required V_{GS} to produce the desired I_D and g_{fs}. Then the values of R_1 and R_2 are chosen to provide the required V_{GS} through voltage divider action. Note that the supply for the R_1–R_2 voltage divider must have a negative voltage to supply the required -1.1V. This is inconvenient since the V_{DD} supply for an N-channel MOSFET is positive.

Another factor that makes the external-bias-only method impractical is the variation in device characteristics. For example, as shown in Fig. 2–14, the drain currents for individual devices may cover a wide range of values, as indicated by the dashed curves (high and low). With a fixed, external-bias of -1.1V, drain current could range from cutoff to 18.5 mA. Some form of dc feedback is obviously desirable to maintain the drain current constant over the normal range of product variation.

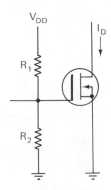

Fig. 2–15 Basic external-bias circuit.

2–4.3 Combined External-Bias and Self-Bias

Figure 2–16 shows the basic circuit for combined external-bias and self-bias. The circuit makes use of a larger value of R_S to narrow the range of drain current to plus or minus a few milliamperes.

Figure 2–17 shows curves of I_{DSS} as a function of I_D for various values of R_S. The normal range of I_{DSS} for the 3N128 is from 5 to 25 mA, or a spread of 20 mA. If the value of R_S is 220 ohms, as calculated for self-bias in Sec. 2–4.1, I_D will range from about 2.5 to 7.5 mA, or a spread of 5 mA. This is an approximate 4–to–1 improvement over the 20 mA spread without any self-bias resistor R_S. Higher values of R_S produce tighter control of the spread of drain-current values.

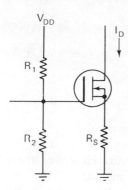

Fig. 2–16 Basic combined external-bias and self-bias circuit.

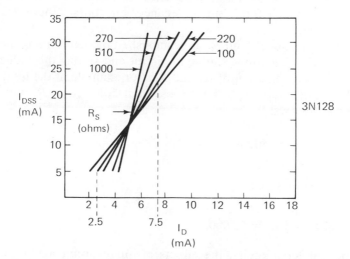

Fig. 2–17 I_{DSS} as a function of I_D for 3N128 with various values of R_S. (Courtesy RCA)

For example, the circuit of Fig. 2–16 may be required to maintain drain current constant within ±1 mA for the same conditions given in the previous example.

Figure 2–17 shows that a value of R_S equal to or greater than 1000 ohms will satisify the required drain-current tolerance. With an R_S of 1000, I_D will range from about 4 to 6 mA. However, a quiescent current of 5 mA through an R_S of 1000 ohms produces a V_{GS} of -5V, which is incompatible with a drain current of 5 mA. Figure 2–14 shows that the device will be cut off, and I_D will be zero with a V_{GS} of -5V. Therefore, an external-bias must be used in conjunction with the self-bias.

The circuit parameters for Fig. 2–16 are calculated using $V_{DS} = 15$V, $I_D = 5$ mA, $V_{GS} = -1.1$V, and $R_S = 1000$, as follows:

$$V_S = I_D R_S = (0.005)(1000) = 5\text{V}$$

$$V_G = V_{GS} + V_S = -1.1 + 5 = 3.9\text{V}$$

$$V_{DD} = V_{DS} + V_S = 15 + 5 = 20\text{V}$$

$$\frac{V_{DD}}{V_G} = \frac{R_1 + R_2}{R_2} = \frac{20}{3.9} = 5.12$$

In theory, any values could be used for R_1 and R_2 provided that the ratio matched the V_{DD}/V_G ratio. However, there are certain practical limitations. First, the input impedance of the circuit is approximately equal to the parallel resistance of R_1 and R_2, or $(R_1 R_2)/(R_1 + R_2)$. If the application requires a specific input impedance for the circuit, R_1 and R_2 can be selected to match the input impedance. Of course, there are upper and lower limits.

The *lower limits* of R_1 and R_2 are established by determining the *maximum permissible loading* of the input circuit and setting this value equal to the parallel combination of R_1 and R_2. For example, if the total shunting of the input circuit is to be no less than 50K ohms, R_1 and R_2 are calculated as follows:

$$\frac{R_1 R_2}{R_1 + R_2} = 50,000 \qquad R_1 \approxeq \left(\frac{V_{DD}}{V_G}\right) \times \text{input impedance}$$

$$\frac{R_1 + R_2}{R_2} = 5.12$$

$R_1 = 256,000$, and $R_2 = 62,000$.

In practical RF circuits, the effects of input circuit loading can frequently be eliminated by the use of the circuit arrangement shown in

Fig. 2–18. Variations of this circuit are discussed throughout the remaining sections of this book.

The *upper limits* of R_1 and R_2 are usually determined by practical considerations of the resistor component values, based on absolute values
of I_{GSS}. Typically, I_{GSS} for a MOSFET is extremely small.

In those unique applications where I_{GSS} is a significant factor, a maximum value for the parallel combination of R_1 and R_2 can be determined
by dividing the total *permissible change in voltage* across the combination by the maximum allowable value of I_{GSS} at the expected operating
temperature, as determined from the datasheet.

Because I_{GSS} consists of leakage currents from both drain and source,
and these currents are usually measured with a maximum-rated voltage
stress on the gate with respect to all other elements, the published value
of I_{GSS} is generally much higher than that which could be expected under
typical circuit conditions. As a result, the values of R_1 and R_2 determined in this manner are conservative.

Another method of setting the value of R_1 and R_2 is to select a value
for R_1 such that I_1 is at least six times greater than the maximum I_{GSS}.
This method is discussed further in Sec. 2–6.

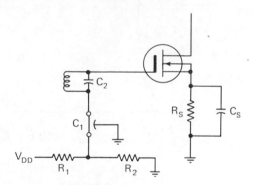

Fig. 2–18 Circuit used to minimize input-circuit loading.

2–4.4 Substrate Biasing

Many single-gate MOSFETs include provisions for separate connection to the substrate. It is sometimes desirable to apply a
separate bias to the substrate, and to use the substrate as an additional
control element. A simple arrangement for achieving this bias is shown in
Fig. 2–19. In this circuit, the substrate bias V_{US} is equal to $I_D (R_1 + R_2)$.
The gate bias voltage V_{GS} is equal to $I_D R_1$.

One application in which substrate bias is essential is the attenuator circuit shown in Fig. 2–20. A MOSFET is very useful as an attenuation device. This is because a MOSFET acts as a fairly linear resistance whose conductivity can be drastically changed by means of a dc voltage applied to the gate. In the circuit of Fig. 2–20, for example, a signal applied to the drain can be attenuated by application of a positive voltage to the MOSFET gate. The attenuation A_V obtained is given by

$$A_V = \frac{R_D}{R_D + R_3}$$

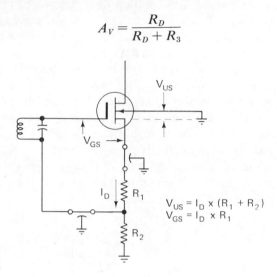

$$V_{US} = I_D \times (R_1 + R_2)$$
$$V_{GS} = I_D \times R_1$$

Fig. 2–19 Basic substrate biasing circuit.

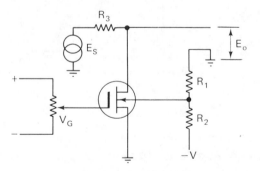

Fig. 2–20 Example of substrate bias circuit used to provide attenuation.

where R_D, the MOSFET channel resistance, is a function of bias voltage V_G and can be varied from approximately 100 ohms to 10^5 megohms.

Because of MOSFET construction, the drain must always be positive with respect to the substrate so that the drain-to-source diode (diffusion) will not be biased into conduction. Therefore, the substrate must be reverse-biased to at least the peak value of the negative-going signal that might be applied to the drain. Figure 2–20 shows how this re-verse-bias is obtained (at the junction of R_1 and R_2, with R_2 connected to a negative supply).

2–4.5 Enhancement Device Biasing

Although the biasing considerations already covered are applicable to all types of single-gate MOSFETs, enhancement-type de-vices *must be turned on* before they can be used as amplifiers. Thus, applied bias such as shown in Figs. 2–15 and 2–16 must always be used with enhancement-type MOSFETs. In addition, it is desirable to narrow the range of drain current by means of a source resistor, such as that shown in Fig. 2–16, that produces self-bias after the transistor is turned on.

As an example of this type of biasing, assume that a *P*-channel en-hancement-type MOSFET is to be operated at room temperature with a supply voltage of 19V, a source resistance of 1000 ohms, and a drain current of 1 mA, as shown in Fig. 2–21. To complete the bias circuit,

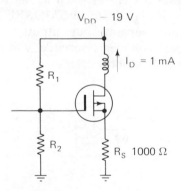

Fig. 2–21 Bias circuit for an enhancement-type MOSFET.

it is necessary to determine the values of R_1 and R_2 to satisfy a total input-loading requirement of $10K$ ohms. Further assume that the MOSFET has a typical threshold voltage of -5.3V, and requires a gate voltage of ap-proximately -9.2V for a drain current of 1 mA. As discussed in Sec. 1–6.1, the threshold voltage $V_{GS(th)}$ (or V_{TH} as it is sometimes described) for an enhancement-type device is comparable to the cutoff voltage $V_{GS(OFF)}$ for a depletion-type device, and is the value of gate voltage re-quired to initiate drain current. Threshold voltage is usually specified for a drain-current value between 10 and 100 mA.

With these established values, the circuit parameters for the network of Fig. 2–21 are then calculated as follows:

$$V_S = I_S R_S = (-0.001)(1000) = -1V$$

$$V_{DS} = V_{DD} - V_S = -19V + 1 = -18V$$

$$V_G = V_{GS} + V_S = -9.2 - 1 = -10.2V$$

$$\frac{R_1 + R_2}{R_2} = \frac{V_{DD}}{V_G} = \frac{19}{10.2} = 1.86$$

$$\frac{R_1 R_2}{R_1 + R_2} = 10,000 \text{ ohms}$$

$$R_1 = 18,600$$

$$R_2 = 21,500$$

Note that a first trial value for R_1 is equal to the ratio of V_{DD}/V_G times the desired input loading, or $1.86 \times 10,000$.

2–4.6 Dual-Gate MOSFET Bias

A dual gate MOSFET such as shown in Fig. 2–22(a) is actually a combination of two single-gate MOSFETs arranged in a cascode configuration (inputs separate but outputs in series), as shown in Fig. 2–22(b). The element voltages associated with each of the individual transistors can be analyzed as follows:

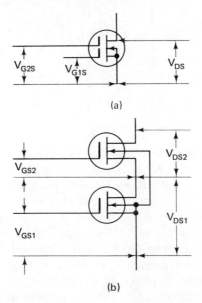

(a)

(b)

Fig. 2–22 Circuits showing element voltage associated with dual-gate MOSFETs.

$$V_{DS} = V_{DS1} + V_{DS2}$$

$$V_{G2S} = V_{DS1} + V_{GS2}$$

$$V_{G1S} = V_{GS1}$$

Curves of the voltage distributions for the 3N140 dual-gate MOSFET are shown in Fig. 2–23. It can be seen that for an applied gate–No. 1–to–source voltage V_{G1S} of zero, a supply voltage V_{DD} of +15V and a gate–No. 2–to–source voltage V_{G2S} of +3V, the actual drain voltage across the grounded-source is approximately +2.75V, and gate No. 2 is 0.25V positive with respect to its own source. These curves explain the logic behind the apparently high positive gate–No. 2 voltage (in the order of +4V) recommended for typical operation of dual-gate MOS-FETs.

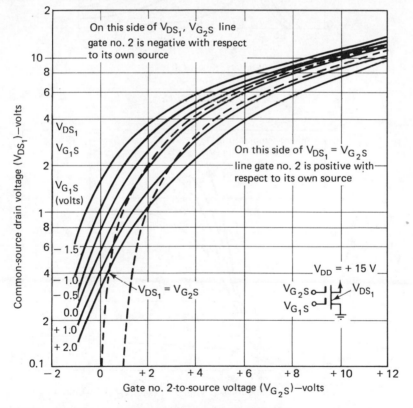

Fig. 2–23 Voltage distributions for the 3N140 dual-gate MOSFET. (Courtesy RCA)

Operating curves for the 3N140 are shown in Fig. 2–24. These curves can be used to establish a quiescent operating condition for the tran-

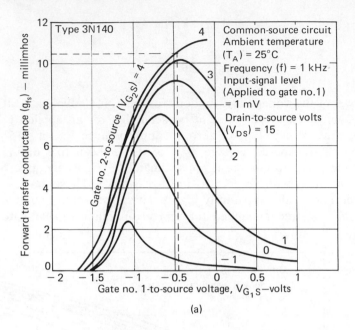

(a)

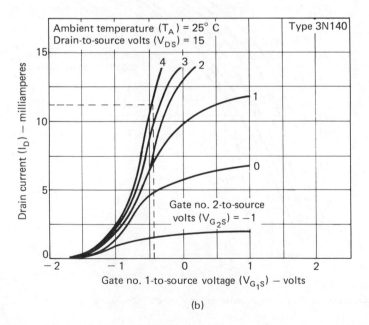

(b)

Fig. 2–24 Characteristic curves for the 3N140. (Courtesy RCA)

sistor. For example, a typical application may require the 3N140 to be operated at a drain-to-source voltage V_{DS} of 15V and a transconductance g_{fs} of 10.5 mmhos. As shown in Fig. 2–24(a), the desired value of g_{fs} can be obtained with a gate–No.2–to–source voltage V_{G2S} of +4V and a gate-No.1-to-source voltage V_{G1S} of −0.45V. From Fig. 2–24(b), the drain current compatible with these gate voltages is 10 mA.

Two biasing arrangements which can be used to provide these operating conditions for the 3N140 are shown in Fig. 2–25. For the previous application, it may be assumed that shunt resistance for gate No. 1 should be $25K$ ohms, and that the dc potential on gate No. 2 should be fixed at RF ground (by means of the feed-thru capacitor). The remaining parameters for the biasing circuits can then be obtained from the curves showing I_D as a function of R_S in Fig. 2–26(a), with $R_S = 270$ ohms:

$$V_S = I_D R_S = +2.7\text{V}$$

$$V_{G1} = V_{G1S} + V_S = (-0.45) + (+2.7) = +2.25\text{V}$$

$$V_{G2} = V_{G2S} + V_S = (+4) + (+2.7) = +6.7\text{V}$$

$$V_{DD} = V_{DS} + V_S = (+15) + (+2.7) = +17.7\text{V}$$

The values of the resistance voltage dividers required to provide the appropriate gate voltages are determined in the same manner as for single-gate MOSFETs (Sec. 2–4.3). For the circuit of Fig. 2–25(a), R_3 is 197,000 ohms, R_4 28,600 ohms, $R_1 = 11,000$, ohms, and $R_2 = 6,700$ ohms.

The circuit of Fig. 2–25(a) is normally used in RF-mixer applications and in RF amplifier circuits that do not use AGC. The circuit of Fig. 2–25(b) is recommended for the application of AGC voltage to RF amplifier stages. In this circuit, the RF signal is applied to gate 1, with the AGC voltage applied to gate 2.

The dual-gate MOSFET is useful in AGC-supplied RF amplifiers since almost no AGC power is required. This is because of the high dc input resistance of a MOSFET.

2–4.7 Bias for Temperature Compensation

Unlike two-junction transistors, MOSFETs show a negative temperature coefficient for typical values of drain current I_D. That is, I_D and dissipation P_D *decrease* as temperature increases. Thus, there is no possibility of I_D runaway (thermal runaway) with elevated temperatures. However, transconductance and RF power gain also decrease as

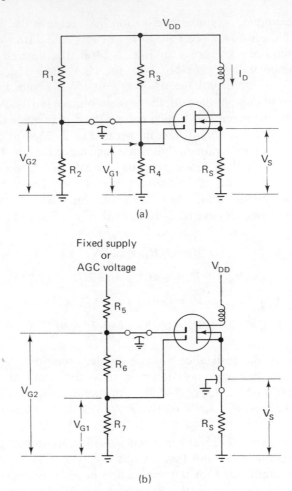

Fig. 2–25 Typical bias circuits for the 3N140. (Courtesy RCA)

temperature increases. Figure 2–26(b) shows curves of drain current and transconductance as a function of temperature. These curves also show the compensating effects produced by the use of source resistance R_S. Note that variations in drain current are reduced significantly by the use of an R_S value of 1000 ohms.

Variations in transconductance can be virtually eliminated by application of a gain-control voltage from a temperature-dependent voltage divider network to gate 2. For example, in a manufacturer's test, the values of the resistance voltage dividers in the circuit of Fig. 2–25(a) were determined to provide a transconductance of 9.5 μmhos at ambient temperature. Then the device temperature was varied through the range from −45 to +100°C. The values of gate 2–to–source voltage V_{G2S} re-

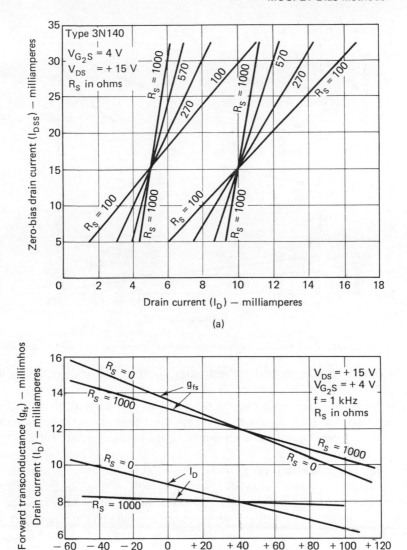

Fig. 2-26 Drain-current and transconductance curves for 3N140. (Courtesy RCA)

quired to maintain a constant transconductance over the entire temperature range, for R_S values of 0 and 1000 ohms, are shown in Fig. 2–27.

In a practical circuit, the required voltages can be applied to gate 2 if R_1, or the combination of R_1 and R_2, is a temperature-sensitive resistor (thermistor) that is thermally linked to the MOSFET package. Gen-

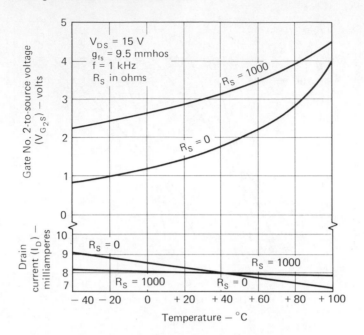

Fig. 2–27 Drain current and gate- 2-to-source voltage for constant I_D as a function of temperature. (Courtesy RCA)

erally, this is accomplished by mounting the thermistor directly on the MOSFET.

The thermistor network can be designed to provide a desired voltage characteristic at gate 2, either to keep the transconductance constant or to permit some variation with temperature to compensate for changes in other stages.

During the manufacturer's tests, the effects on other stages were measured on a 3N140 MOSFET in the circuit of Fig. 2–25(a). Drain current was 8 mA, frequency was 200 MHz, and temperature was varied from 0 to 100°C. The effects of temperature given in percentages on other stages may be summarized as follows: R_{in}, C_{in}, C_{out}, and $C_{feedback}$ all 1 percent, $R_{out} + 45$ percent.

2–4.8 Summary of MOSFET Biasing

All MOSFETs can be biased similarly. Uniform quiescent operating points can be easily achieved in MOSFETs by using circuit designs that include a source resistance R_S. For a given I_{DSS} range, the value of the source resistance inversely affects the in-circuit I_D spread. An increase in the value of the source resistance minimizes variations in I_D as a function of temperature. The dual-gate MOSFET is ideally suited for use in gain-controlled stages. Dual-gate MOSFET biasing can provide various types of AGC action, including temperature compensation to assure constant output.

2–5. DESIGNING THE BASIC MOSFET
BIAS NETWORK

The following example is presented to show how the transfer characteristic curves from a datasheet are used in the design of a typical MOSFET bias network. The circuit involved is shown in Fig. 2–28. Note that both external (or fixed-bias) and self-bias are used.

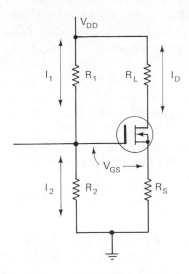

Fig. 2–28 The basic MOSFET stage.

As discussed later in this chapter, the basic bias network can be modified to produce a MOSFET amplifier stage with desired characteristics (stage gain, input/output impedance, etc.).

Keep one point in mind when studying the following bias scheme: The purpose of the basic bias circuit is to establish a given I_D, and to maintain that I_D (plus or minus some given tolerance) over a given temperature range. Generally, this is to keep the transconductance at some given level. In the following discussion, no particular consideration is given to stage characteristics (operating point, gain, impedance, etc.). Instead, we concentrate on how to set the I_D, and how to keep the I_D constant.

Of course, since MOSFETs rarely operate under static conditions, the basic bias circuit is used as a reference or staring point for design. The actual circuit configuration, and especially the bias circuit values, should be selected on the basis of dynamic circuit conditions (desired output voltage swing, expected input signal level, etc.). For example, if a MOSFET is used as a linear amplifier, the output (drain terminal) should be at one-half the supply voltage (for a typical stage). This permits maximum voltage swing. As a result, the value of R_L should be selected to provide the voltage drop equal to one-half the supply.

2–5.1 Design Example

Assume that the circuit of Fig. 2–28 is to maintain I_D at 1 ± 0.025 mA, over a temperature range from -55 to $+125°C$, with a supply voltage V_{DD} of 30V, using a MOSFET with transfer characteristics similar to those of Fig. 2–29 (N-channel depletion type).

The first step is to draw a one/R_S load line on the transfer characteristics as shown in Fig. 2–29. As illustrated by the equations, the value of R_S is set by the limits of V_{GS} and I_D. The value of $V_{GS(min)}$ is the point where the $I_{D(min)}$ of 0.75 mA crosses the high-temperature limit curve of $+125°C$, or approximately 0.8V. The value of $V_{GS(max)}$ is the point where the $I_{D(max)}$ of 1.25 mA crosses the low-temperature limit curve of $-55°C$, or approximately 1.9V.

Using these values, the first trial value for R_S is:

$$R_S = \frac{1.9 - 0.8}{1.25 - 0.75} = 2.2\text{K ohms}$$

The fixed-bias voltage V_G is determined from the intercept of the $1/R_S$ load line with the V_{GS} axis, and is computed by using the same set of values shown in the Fig. 2–29 equations:

$$V_G = \frac{(0.75 \times 1.9) - (1.25 \times 0.8)}{0.5} = 0.85\text{V}$$

The maximum value of R_1 is determined by the maximum value of I_{GSS}. That is, the value of I_{GSS} at the highest temperature involved. Assume that the MOSFET has a maximum I_{GSS} of 5 nA at $+125°C$. The variation in V_G versus temperature will not be too great if a value for R_1 is chosen so that I_1 is at least six times greater than the maximum I_{GSS}. For extra stability, let I_1 equal ten times maximum I_{GSS}, or 10×5 nA $= 50$ nA.

The drop across R_1 is equal to $V_{DD} - V_G$, or $30 - 0.85 = 29.15$V. With an I_1 of 50 nA and a drop across R_1 of 29.15V, the approximate value of R_1 is 29.15V/50 nA = 583 megohms.

With the value of R_1 established at 583 megohms, the value of R_2 is found from a simple voltage divider relationship as shown in the equations in Fig. 2–29:

$$R_2 = \frac{(0.85) \times (583 \times 10^6)}{30 - 0.85} \sim 17 \text{ megohms}$$

The value of R_L is determined by desired circuit performance, or output characteristics, or both. For example, an increase in the value of R_L increases gain, but decreases stability. Likewise, the value of R_L sets the approximate output impedance of the circuit (when the MOSFET is connected as a common source amplifier). These factors are discussed in Sec. 2–6.

With all of the bias resistance values established, the circuit of Fig. 2–28 can be converted to a single MOSFET amplifier stage by the addition of input and output coupling capacitors. This conversion is described in Sec. 2–6.

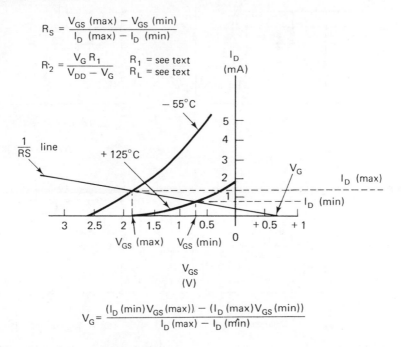

Fig. 2–29 1/RS load line for basic MOSFET stage.

2-6. DESIGNING THE BASIC MOSFET
LINEAR AMPLIFIER

Figure 2–30 shows three basic single-stage amplifier configurations that use dual-gate-protected MOSFETs as triodes and as
tetrodes in common-source, common-drain, and common-gate circuits.

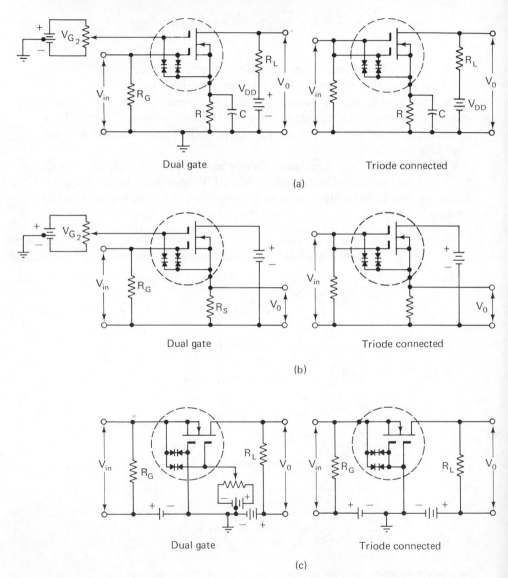

Fig. 2–30 Three theoretical single-stage amplifier configurations using dual-gate MOSFETS. (Courtesy RCA)

Each configuration has its own particular advantages for specific applications. The dual-gate device has an added advantage in any of these configurations in that gate 2 provides (a) reduced gate-to-drain capacity (by at least one-half), and (b) a convenient means for controlling gain of the stage by adjusting the dc potential applied to gate 2.

All the circuits in Fig. 2–30 are theoretical. In this section, we shall discuss practical versions of the circuits. However, before going into the complete MOSFET circuits, we shall discuss the small-signal characteristics of the basic MOSFET stage.

2–6.1 Small-Signal Analysis of Basic MOSFET Stage

Small-signal analysis of a MOSFET amplifier stage is easily accomplished with reasonable accuracy by using a few simple equations. The MOSFET model for the analysis is shown in Fig. 2–31. Note that this model omits all capacitance, including C_{rss}. By omitting all capacitance, and using only the real parts of Y_{is}, Y_{fs} (or g_{fs}) and Y_{os}, the model and the accompanying equations are useful up to about 100 kHz. If capacitance effects are included, the equations are useful up to several megahertz. The effects of capacitance on RF amplifiers are discussed in Chap. 3.

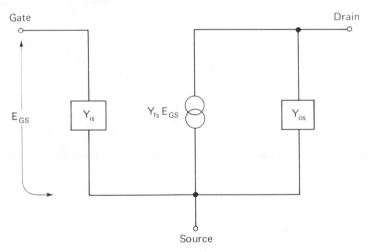

Fig. 2–31 Small-signal analysis of basic MOSFET stage (omitting all capacitance).

Figures 2–32, 2–33, and 2–34 show schematics for common-source, common-drain (source-follower), and common-gate circuits. In addition, the *approximate* equations for voltage gain, input impedance, and output impedance are included. The equations for finding the resistance values of R_1, R_2, R_S and R_L are discussed in Sec. 2–5. Note that the approximate equations are based on certain assumptions given in the corresponding equations.

The circuits of Figs. 2–32 through 2–34 are for N-channel devices. Simply reverse the polarity for P-channel MOSFETs. All MOSFETs may not conform exactly to the relations shown, but are sufficiently close for first trial values.

For the common-source circuit (Fig. 2–32), the omission of C_{rss} will affect the equations for input impedance and voltage gain. However, for low frequencies (below about 100 kHz) the error is minimal. If *only dc feedback* is required, then the source resistor R_S is bypassed. With a capacitor across R_S, the effects on the design equations are to set R_S at zero. Under these conditions, voltage gain is the product of R_L and Y_{fs}. (Note that g_{fs} can be substituted in any equation of Figs. 2–32 through 2–34.)

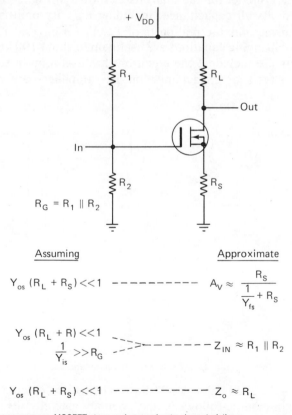

Assuming

$$Y_{os}(R_L + R_S) \ll 1 \quad \text{--------} \quad A_V \approx \frac{R_S}{\dfrac{1}{Y_{fs}} + R_S}$$

Approximate

$$Y_{os}(R_L + R) \ll 1$$
$$\frac{1}{Y_{is}} \gg R_G \quad \text{- - - - - -} \quad Z_{IN} \approx R_1 \| R_2$$

$$Y_{os}(R_L + R_S) \ll 1 \quad \text{-----------} \quad Z_o \approx R_L$$

Fig. 2–32 Basic common-source MOSFET stage and approximate characteristics.

With R_S unbypassed, the circuit characteristics are virtually independent of MOSFET parameters (with the exception of Y_{fs}). Instead, the circuit characteristics (impedances, gain, etc.) are dependent on R_L and R_S. By using precision resistors with close temperature coefficients, the common-source circuit can be made very stable over a wide temperature range.

The common-drain (source-follower) configuration (Fig. 2–33) is a very useful basic circuit. Some of its properties are: A voltage gain always less than unity with no phase inversion; low output impedance (essentially set by the value of R_S), high input impedance, large signal swing, and active impedance transformation.

The common-gate stage (Fig. 2–34) offers impedance transformation opposite that of the source follower. Common-gate produces low input impedance and high output impedance. The voltage gain is less than unity, with no phase inversion.

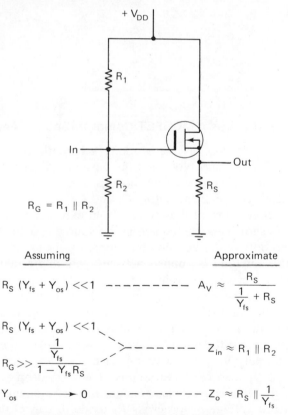

Fig. 2–33 Basic common-drain MOSFET stage and approximate characteristics.

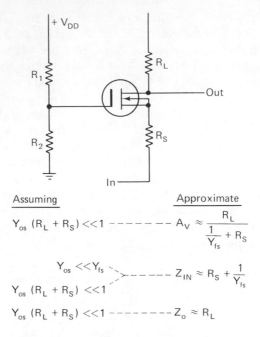

Assuming $\qquad\qquad\qquad\qquad$ Approximate

$$Y_{os}(R_L + R_S) \ll 1 \;\text{-------}\; A_V \approx \dfrac{R_L}{\dfrac{1}{Y_{fs}} + R_S}$$

$$\left.\begin{array}{c} Y_{os} \ll Y_{fs} \\[2pt] Y_{os}(R_L + R_S) \ll 1 \end{array}\right\} \text{-------}\; Z_{IN} \approx R_S + \dfrac{1}{Y_{fs}}$$

$$Y_{os}(R_L + R_S) \ll 1 \;\text{--------}\; Z_o \approx R_L$$

Fig. 2–34 Basic common-gate MOSFET stage and approximate characteristics.

2–6.2 Basic MOSFET Common-Source Amplifier

Figure 2–35 is the working schematic of a basic, single-stage MOSFET amplifier. Note that the basic amplifier circuit is similar to the basic common-source circuit of Figs. 2–28 and 2–32, except that input and output coupling capacitors C_1 and C_2 are added. These capacitors prevent direct-current flow to and from external circuits. A bypass capacitor C_3 is shown connected across the source resistor R_S. Capacitor C_3 is required only under certain conditions, as discussed in Sec. 2–6.3.

Input to the amplifier is applied between gate and ground, across R_2. Output is taken across the drain and ground. The input signal adds to, or subtracts from, the bias voltage across R_2. Variations in bias voltage cause corresponding variations in I_D, and the voltage drop across R_L. Therefore, the drain voltage (or circuit output) follows the input signal waveform except that the output is inverted in phase. (If the input swings positive, the output swings negative, and vice versa.)

Variations in I_D also cause variations in voltage drop across R_S, and a change in the gate-source bias relationship. The change in bias that results from the voltage drop across R_S tends to cancel the initial bias change caused by the input signal, and serves as a form of *negative feedback* to increase stability (and limit gain). This form of gate-source feed-

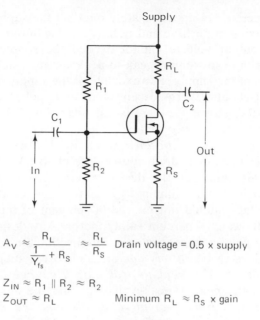

$$A_V \approx \frac{R_L}{\dfrac{1}{Y_{fs}} + R_S} \approx \frac{R_L}{R_S} \quad \text{Drain voltage} = 0.5 \times \text{supply}$$

$$Z_{IN} \approx R_1 \parallel R_2 \approx R_2$$
$$Z_{OUT} \approx R_L \qquad\qquad \text{Minimum } R_L \approx R_S \times \text{gain}$$

Fig. 2–35 Basic common-source MOSFET amplifier stage.

back is known as *stage feedback* or *local feedback*, since only one stage is involved. As discussed in later sections, *overall feedback* or *loop feedback* is sometimes used where several stages are involved.

The outstanding characteristic of the circuit in Fig. 2–35 is that circuit characteristics (gain, stability, impedance) are determined primarily by circuit values, rather than by MOSFET characteristics.

In some MOSFET circuits it is possible to omit capacitors C_1 and C_2, because the gate input to a MOSFET acts as a capacitor. Generally, the need for coupling capacitors is set by other stages. For example, the drain or output terminal of the MOSFET is at some dc voltage with respect to ground. If the stage following the MOSFET cannot tolerate this dc voltage, an output coupling capacitor (C_2) must be used. The same is true of input capacitor C_1. The problem of coupling capacitors in MOSFET amplifiers is discussed further in Sec. 2–6.9.

2–6.2.1 Design Considerations

The circuit shown in Fig. 2–35 uses an *N*-channel MOS-FET. Reverse the power supply polarity if a *P*-channel MOSFET is used.

If a *maximum supply voltage is specified* in the design problem, the maximum peak-to-peak output voltage is set. For class *A* operation, the drain is operated at approximately one-half the supply voltage, permitting the maximum positive and negative swing of output voltage. The peak-to-peak output voltage cannot exceed the source voltage. Generally, the absolute maximum peak-to-peak output can be between 90 and 95 percent of the supply. For example, if the supply is 20V, the drain will operate at about 10V (quiescent point or Q-point), and swing from about 1V to 19V. However, there is less distortion if the output is one-half to one-third of the supply voltage.

If a supply voltage is not specified, two major factors should determine the value: The maximum drain-source V_{DS} of the MOSFET, and the desired output voltage (or the desired drain voltage at the operating point). Obviously, the maximum V_{DS} cannot be exceeded. Preferably, the supply voltage should not exceed 90 percent of the maximum V_{DS} rating. This allows a 10 percent safety factor. Any desired output voltage (or drain Q-point voltage) can be selected within these limits.

If the circuit is to be *battery-operated*, choose a supply voltage that is a multiple of 1.5. If a *peak-to-peak output voltage is specified*, add 10 percent (to the peak-to-peak value) to find the absolute minimum supply voltage. If a *drain Q-point voltage is specified*, double the drain Q-point voltage. If *minimum distortion* is required, use a supply that is two to three times the desired output voltage. *If the input and/or output impedances are specified*, the resistance values (R_1, R_2, R_L, R_S) are set, as shown in Fig. 2–36. However, there are certain limitations for R_2 and R_L imposed by tradeoffs (for gain, impedance match, operating point, etc.).

For example, the output impedance is set by R_L. If R_L is increased to match a given impedance, the gain will increase (all other factors remaining equal). However, an increase in R_L will lower the drain voltage Q-point, since the same amount of I_D will flow through R_L and produce a larger voltage drop. This reduces the possible output voltage swing. A reduction in R_L has the opposite effect; increasing the drain voltage Q-point, but still reducing output voltage swing.

When R_1 is much larger than R_2 (which is generally the case), the input impedance of the circuit is set by the value of R_2. If R_2 is increased (or decreased) far from the value found in Sec. 2–5, the no-signal I_D point will change. For example, in Sec. 2–5, the I_D is set at 1 mA. If this value is selected to provide a given forward transconductance, and R_2 is changed drastically, the drain current will change, and the MOSFET will no longer provide the same transconductance. Generally, this is the least desirable alternative. Typically, the common-source MOSFET circuit is chosen for its high input impedance, thus presenting a low current drain to the preceding circuit. If the MOSFET stage must provide a low input impedance, the common-gate circuit of Fig. 2–34 is generally preferred.

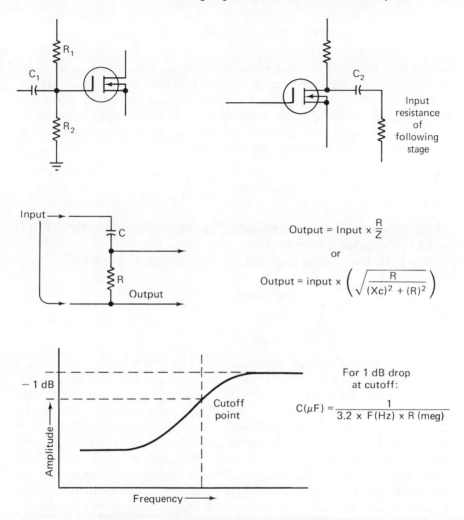

$$\text{Output} = \text{Input} \times \frac{R}{Z}$$

or

$$\text{Output} = \text{input} \times \left(\sqrt{\frac{R}{(X_c)^2 + (R)^2}} \right)$$

For 1 dB drop
at cutoff:

$$C(\mu F) = \frac{1}{3.2 \times F(Hz) \times R \text{ (meg)}}$$

Fig. 2–36 Formation of high-pass RC filter by coupling capacitors and related resistances.

If used, the values of C_1 and C_2 are dependent on the low-frequency limit at which the amplifier is to operate. As frequency increases, capacitive reactance decreases and the coupling capacitor becomes (in effect) a short to the input signal. Thus, the high-frequency limit need not be considered in audio circuits, but is of consequence in RF circuits (as discussed in Chap. 3). Capacitor C_1 forms a high-pass RC filter with R_2. Capacitor C_2 forms another high-pass filter with the input resistance of the following stage (or the load). This condition is shown in Fig. 2–36. The input voltage is applied across the capacitor and resistor in series. The output is taken across the resistance. The relation of input voltage to output voltage is:

$$\text{output voltage} = \text{input voltage} \times \frac{R}{Z}$$

where R is the dc resistance value, and Z is the impedance obtained by the vector combination of series capacitive reactance and dc resistance.

When the reactance drops to approximately one-half of the resistance, the output drops to approximately 90 percent of the input (or approximately 1 dB loss). Using the 1–dB loss as the low-frequency cutoff point, the value of C_1 or C_2 can be found by:

$$\text{Capacitance} = \frac{1}{(3.2 \; FR)}$$

where capacitance is in microfarads, F is the low-frequency limit in Hz, and R is resistance in megohms.

If a 3–dB loss can be tolerated, the capacitance is found by:

$$\text{capacitance} = \frac{1}{6.2 \; FR}$$

2–6.2.2 *Design Example*

Assume that the circuit of Fig. 2–35 is to be used as a single-stage voltage amplifier. The desired output is a minimum of 8V (peak-to-peak) with an input of 1V. This requires a gain of at least 8. The low-frequency limit is 30 Hz, with a high-frequency limit of 100 kHz. Minimum distortion is desired. (The circuit should not be overdriven.) The supply voltage is specified as +30V. The MOSFET selected has a minimum Y_{fs} (or g_{fs}) of 5000 micromhos with an I_D of 1 mA. Maximum Y_{fs} is 10,000 micromhos. Input and output impedances are not specified, but must be calculated for reference to other circuits. Coupling capacitors C_1 and C_2 must be used, due to voltages of external circuits. The MOSFET used has the same characteristics as described in Sec. 2–5.1.

calculate resistance values—The first step is to calculate the resistance values as described in Sec. 2–5.1. Since both problems require an I_D of 1 mA, the values are the same for R_1 (583M), R_2 (17M), and R_S (2.2K).

supply voltage and operating point—With a supply voltage of 30V and an output of 8V, the ratio is much greater than the required three-to-one (or two-to-one), so distortion should be at a minimum. With an 8V peak-to-peak output, the operating point (drain voltage) can be anything from about 5 to 25V. However, the value of R_L has an effect on both gain and operating point. A larger value of R_L will increase gain. To find the minimum value of R_L, multiply the value of R_S by the minimum gain, or

2.2K $\times$ 8 = 17.6K. A value of 20K for R_L should produce the desired gain, and will produce an operating point of 10V (20K $\times$ 1 mA = 20V; 30V $-$ 20V = 10V).

minimum gain — Since a minimum gain of 8 is specified, use the minimum value of Y_{fs} (5000 μmhos) to find gain. Using the longer equation of Fig. 2–35, the gain is:

$$\text{Gain} \approx \frac{20K}{(1/5000 \ \mu\text{mhos}) + 2.2K} = \frac{20K}{0.2K + 2.2K}$$

$$= \frac{20K}{2.4K} \approx 8.33$$

maximum gain — With a possible Y_{fs} of 10,000 μmhos, the maximum gain is:

$$\text{Gain} \approx \frac{20K}{(1/10,000 \ \mu\text{mhos}) + 2.2K} \approx \frac{20K}{2.3K} \approx 8.7$$

Thus, with a maximum 1V input, the maximum output is approximately 8.7V. This can easily be handled by the 10V operating point.

Note that the equations of Fig. 2–35 show that approximate voltage gain is equal to R_L/R_S, without regard to Y_{fs}. This simplified rule holds true unless Y_{fs} and R_S are very small (in relation to R_L). Thus, for a quick check of gain, use the R_L/R_S equation. If greater accuracy is desired, use the longer equation.

input impedance — Input impedance of the Fig. 2–35 circuit is the parallel combination of R_1 and R_2. However, since R_1 is many times (more than 10) the value of R_2, the approximate circuit input impedance is equal to R_2, or about 17M.

output impedance — Output impedance of the Fig. 2–35 circuit is approximately equal to R_L, or 20K.

coupling capacitors — The value of C_1 forms a high-pass filter with R_2. The high limit of 100 kHz can be ignored. The low-frequency limit of 30 Hz requires a capacitance value of:

$$C_1 \approx \frac{1}{3.2 \times 30 \times 17} \approx 0.0006 \ \mu\text{F}$$

This will provide an approximate 1–dB drop at the low-frequency limit of 30 Hz. If a greater drop can be tolerated, the capacitance value of C_1 can be lowered. The value of C_2 is found in the same manner, except the resistance value must be the load resistance. The voltage values of C_1 and C_2 should be 1.5 times the maximum voltage involved, or 30 $\times 1.5 = 45$V.

sufficient feedback—As a final check of design values, compare the calculated gain versus the ratio of R_L/R_S. The gain should be at least 75 percent of the resistance ratio. If so, there is sufficient feedback to be of practical value. In this example, the ratio is approximately 9, with the gain slightly over 8.3; 75 percent of 9 is approximately 6.75. Thus, the gain is greater, and there is sufficient feedback.

2–6.3 Basic MOSFET with Source Resistance Bypass

Figure 2–35 shows (in phantom) a bypass capacitor C_3 across source resistor R_S. This arrangement permits R_S to be removed from the circuit as far as the signal is concerned, but leaves R_S in the circuit (in regard to direct current). With R_S removed from the signal path, the voltage gain is approximately equal to $Y_{fs} \times R_L$. Thus, the use of a bypass capacitor permits a temperature-stable dc circuit to remain intact, while providing a high signal gain.

2–6.3.1 *Design Considerations*

A source resistance bypass capacitor also creates some problems. The Y_{fs} changes with frequency, and from MOSFET to MOSFET. Thus, circuit gain can only be approximated. The source bypass is recommended where maximum voltage gain must be obtained from a single stage.

The value of C_3 can be found by:

$$\text{capacitance} = \frac{1}{6.2F(R_S \times 0.2)}$$

where capacitance is in microfarads, F is low-frequency limit in Hz, and R_S is in megohms.

2–6.3.2 *Design Example*

Assume that C_3 is to be used as a source bypass for the circuit described in the previous design example (Sec. 2–6.2) to increase voltage gain. All the circuit values remain the same, as does the low-frequency limit of 30 Hz. Assume that the MOSFET has a Y_{fs} of 5000 μmhs minimum (same as previous example), and that the same 8V minimum output is required.

When C_3 is added, the equation for approximate voltage gain changes to:

$$\text{Gain} \approx Y_{fs} \times R_L \quad \text{or} \quad 0.005 \times 20\text{K} \approx 100$$

With this gain, the input can be reduced to about 80 mV (from the 1V in the previous example) to produce the same 8V output. Keep in mind that the same MOSFET might have a Y_{fs} of 10,000 μmhos (as stated in the previous example). If so, the stage voltage gain will be 200, and the 8V output can be accomplished with about 40 mV. Likewise, if the 80 mV input is applied to the circuit with a 10,000 μmhos Y_{fs} MOSFET, the output will be over 8V. This output may or may not produce distortion or clipping with a Q-point of 10V. However, the output may be excessive for the load or stage following the circuit. These factors must be considered using source bypass.

The low-frequency limit of 30 Hz requires a C_3 capacitance value of:

$$C_3 \approx \frac{1}{6.2 \times 30 \times (0.0022 \times 0.2)} \approx 12 \ \mu F$$

This value provides a reactance across R_S that is about one-fifth of R_S, and effectively shorts the source (signal path) to ground. The voltage value of C_3 should be 1.5 times the maximum voltage involved, or 45V.

2–6.4 Basic MOSFET with Partially Bypassed Source

Figure 2–37 is the working schematic of a basic, single-

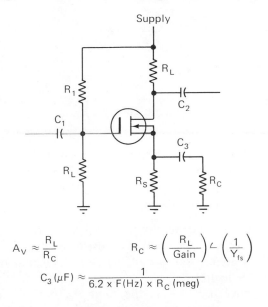

$$A_V \approx \frac{R_L}{R_C} \qquad R_C \approx \left(\frac{R_L}{Gain}\right) \angle \left(\frac{1}{Y_{fs}}\right)$$

$$C_3 \ (\mu F) \approx \frac{1}{6.2 \times F(Hz) \times R_C \ (meg)}$$

Fig. 2–37 Basic MOSFET audio amplifier with partially bypassed source resistor.

stage MOSFET amplifier with a partially bypassed source resistor. This design is a compromise between the basic design without bypass (Sec. 2–6.2) and the fully bypassed source (Sec. 2–6.3). The direct-current characteristics of both the unbypassed and partially bypassed circuits are essentially the same. All circuit values (except C_3 and R_C) can be calculated in the same way for both circuits.

As shown in Fig. 2–37, the voltage gain for a partially bypassed MOSFET amplifier is greater than the unbypassed circuit, but less than that of the fully bypassed circuit. However, the gain can be set to an approximate value by the selection of circuit values (unlike the fully bypassed circuit where gain is entirely subject to variations in Y_{fs}).

2–6.4.1 *Design Considerations*

Design considerations for the circuit of Fig. 2–37 are the same as those for Fig. 2–35 (Sec. 2–6.2), except for the effect of C_3 and R_C.

The value of R_C is chosen on the basis of desired voltage gain. R_C should be substantially smaller than R_S, otherwise there will be no advantage to the partially bypassed design. As shown by the equations, voltage gain is approximately equal to R_L/R_C. This holds true unless both Y_{fs} and R_C are very low (where $1/Y_{fs}$ is about equal to R_C). In such a case, a more accurate gain approximation is:

$$\frac{R_L}{1/Y_{fs} + R_C}$$

The value of C_3 is found by:

$$\text{capacitance} = \frac{1}{6.2FR_C}$$

where capacitance is in microfarads, F is the low-frequency limit in Hz, and R_C is in megohms.

2–6.4.2 *Design Example*

Assume that the circuit of Fig. 2–37 is to be used in place of the Fig. 2–35 circuit, and that the desired voltage gain is 50 (more than that of the unbypassed circuit, but less than that of the fully bypassed circuit). Selection of the component values, supply voltages, operating point, and the like, is the same for both circuits; thus, the only difference in design is selection of values for C_3 and R_C.

The value of R_C is the value of R_L divided by the desired gain, less the reciprocal of minimum Y_{fs}, or: minimum $Y_{fs} = 5000$ μmhos (0.005 mho); the reciprocal of $Y_{fs} = 1/0.005 = 200$; $R_L/\text{gain} = 20K/50 = 400$; $400 - 200 = 200$; $R_C = 200$ ohms.

The low frequency limit of 30 Hz requires a C_3 capacitance value of:

$$\frac{1}{6.2 \times 30 \times 0.0002} \approx 27 \ \mu F$$

The voltage value of C_3 should be 1.5 times the maximum voltage involved, or $45V$.

2–6.5 Basic MOSFET Source Follower (Common Drain)

Figure 2–38 is the working schematic of a basic single-stage MOSFET source-follower (common-drain) circuit. Note that this circuit is similar to that of Fig. 2–33, except that input and output coupling capacitors C_1 and C_2 are added. These capacitors prevent direct-current flow to and from external circuits.

Input to the source follower is applied between gate and ground across R_2. Output is taken across the source and ground. The input signal adds to, or subtracts from, the bias voltage across R_S. Variations in bias voltage cause corresponding variations in I_D and the voltage drop across R_S. Therefore, the source voltage (or circuit output) follows the input signal waveform, and remains in phase.

Variations in voltage drop across R_S change the gate-source bias relationship. This change in bias tends to cancel the initial bias change caused by the input signal and serves as a form of negative feedback to increase stability and limit gain.

The circuit of Fig. 2–38 is used primarily where high input impedance and output impedance (with no phase inversion) are required, but no gain is needed. The source follower is the MOSFET equivalent of the bipolar transistor emitter follower, and the vacuum tube cathode follower.

2–6.5.1 Design Considerations

The design considerations for the source follower of Fig. 2–38 are essentially the same as for the MOSFET common-source amplifier of Sec. 2–6.2, with the following exceptions.

output Q-point — The Q-point voltage at the circuit output (source terminal) is set by I_D under no-signal conditions, and the value of R_S. Since R_S is typically small, the Q-point voltage is quite low in comparison to the common-source amplifier. In turn, the maximum allowable peak-to-peak output voltage is also low. For example, if the source is at 1V with no signal, the maximum possible peak-to-peak output is less than 1V. Of course, the value of R_S can be increased as necessary to permit a higher output.

If the input and/or output impedances are specified in a design problem, the resistance values (R_1, R_2, R_S) are set, as shown in Fig. 2–38. However, there are certain limitations for R_S imposed by tradeoffs (for impedance match and output Q-point).

For example, the output impedance is the parallel resistance combination of R_S and $1/Y_{fs}$. If R_S is made very small (less than 10 times) in relation to $1/Y_{fs}$, then the output impedance is approximately equal to R_S. As discussed, a low value of R_S decreases the source (output) voltage Q-point, thus reducing output voltage swing. If R_S is made large in relation to $1/Y_{fs}$, then the output impedance is approximately equal to $1/Y_{fs}$, and is subject to variation with frequency, and from device to device.

current gain—As shown by the equation of Fig. 2–38, there is no

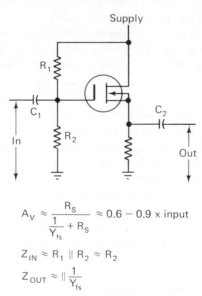

$$A_V \approx \frac{R_S}{\dfrac{1}{Y_{fs}} + R_S} \approx 0.6 - 0.9 \times \text{input}$$

$$Z_{IN} \approx R_1 \parallel R_2 \approx R_2$$

$$Z_{OUT} \approx \parallel \frac{1}{Y_{fs}}$$

Fig. 2–38 Basic common-drain (source follower) MOSFET amplifier stage.

voltage for a source follower. Typically, the output voltage is about 0.6 to 0.9 times the input voltage, depending on the ratio of $1/Y_{fs}$ to R_S. However, the source follower is capable of current gain, and thus of *power gain*. For example, assume that 1V is applied at the input and

0.6V is taken from the output. Further assume that the input impedance is 17M, and the output impedance is 180 ohms. The input power is approximately 0.06 μW, while the output power is about 2 mW, indicating a power gain of over 33,000.

2–6.5.2 Design Example

Assume that the circuit of Fig. 2–38 is to be used as a single-stage source follower. The desired output is a minimum of 0.6V (peak-to-peak) with an input of 1V. The low-frequency limit is 30 Hz. All other conditions are the same as for the design example in Sec. 2–6.2.

calculate resistance values—The first step is to calculate the resistance values as described in Sec. 2–5 and 2–6.2, omitting the calculations for R_L. Since all factors are the same, the values are $R_1 = 587M$, $R_2 = 17M$, $R_S = 2.2K$.

operating point—With an I_D of 1 mA, and an R_S of 2.2K, the drop across R_S (and the operating point) is about 2.2V. This source voltage Q-point will easily permit a 0.6-V peak-to-peak output swing.

minimum output signal voltage—Since a minimum output of 0.6V is specified with a 1V input, the stage loss factor must be no greater than 0.6. Use the minimum value of Y_{fs} (5000 μmhos) to find the loss factor. Using the equation in Fig. 2–38, the loss factor is:

$$\frac{2200}{\dfrac{1}{0.005} + 2200} \approx \frac{2200}{2400} \approx 0.9$$

Thus, the output signal voltage should be about 0.9V peak-to-peak.

input impedance—The input impedance of the Fig. 2–38 circuit is the parallel combination of R_1 and R_2. However, since R_1 is many times (more than 10) the value of R_2, the approximate circuit input impedance is equal to R_2, or about 300K.

output impedance—Output impedance of the Fig. 2–38 circuit is the parallel combination of R_S and $1/Y_{fs}$, or

$$\frac{2200 \times 200}{2200 + 200} \approx 180 \text{ ohms}$$

coupling capacitors—The values of the coupling capacitors C_1 and C_2 are the same as for the example of Sec. 2–6.2.

2-6.6 Basic MOSFET Common Gate

Figure 2–39 is the working schematic of a basic, single-stage MOSFET common-gate amplifier. Note that the circuit is similar to the basic common-gate circuit of Fig. 2–34, except that input and output coupling capacitors C_1 and C_2 are added. These capacitors prevent direct-current flow to and from external circuits.

Input to the common-gate amplifier is applied at the source across a portion of R_S. Typically, the value of R_{S1} is equal to R_{S2}, although it may be necessary to divide the resistance value unequally. In any event, the total value of R_S ($R_{S1} + R_{S2}$) must be considered when calculating the direct-current characteristics of the circuit. Output is taken across the drain and ground. The input signal adds to, or subtracts from, the bias voltage across R_S. Variations in bias voltage cause corresponding variations in I_D and the voltage drop across R_L. The drain voltage (or circuit output) follows the input signal in phase.

Variations in voltage drop across R_S change the gate-source bias relationship. This change in bias tends to cancel the initial bias change caused by the input signal and serves as a form of negative feedback to increase stability (and to limit gain).

The circuit of Fig. 2–39 is used primarily where low input impedance

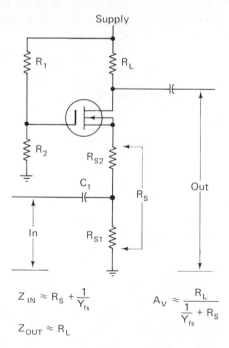

$$Z_{IN} \approx R_S + \frac{1}{Y_{fs}}$$

$$A_V \approx \frac{R_L}{\frac{1}{Y_{fs}} + R_S}$$

$$Z_{OUT} \approx R_L$$

Fig. 2–39 Basic common-gate MOSFET amplifier stage.

and high output impedance (with no phase inversion) are required. Gain is determined primarily by circuit values rather than by MOSFET characteristics. The common-gate amplifier is the MOSFET equivalent of the bipolar transistor common-base amplifier and the vacuum tube common-grid amplifier.

2–6.6.1 *Design Considerations*

Design considerations for the common-gate amplifier of Fig. 2–39 are essentially the same as for the MOSFET common-source amplifier of Sec. 2–6.2, with the following exceptions.

If the input and/or output impedances are specified in a design problem, the resistance values (R_1, R_2, R_L, R_S) are set, as shown in Fig. 2–39. However, note that input impedance of the circuit is dependent on the reciprocal of the Y_{fs} ($1/Y_{fs}$) factor. This is true unless the value of R_S is many times (at least 10) that of $1/Y_{fs}$.

The values of coupling capacitors C_1 and C_2 are dependent on the low-frequency limit at which the amplifier is to operate. Capacitor C_1 forms a high-pass RC filter with R_{S2}. Capacitor C_2 forms another high-pass filter with the input resistance of the following stage (or the load).

Using a 1 dB loss as the low-frequency cutoff point, the value of C_1 can be found by:

$$\text{capacitance} = \frac{1}{3.2 \, F R_{S2}}$$

where capacitance is in microfarads, F is the low-frequency limit in Hz, and R_{S2} is in megohms.

2–6.6.2 *Design Example*

Assume that the circuit of Fig. 2–39 is to be used as a single-stage common-gate amplifier. All conditions are the same as for the design example in Sec. 2–6.2, with the following exceptions.

calculate resistance values—The first step is to calculate resistance values as described in Sec. 2–5 and 2–6.2. Since all factors are the same, the values are: $R_1 = 583M$, $R_2 = 17M$, $R_L = 20K$, $R_S = 2.2K$.

input impedance—Input impedance of the Fig. 2–39 circuit is the combination of $R_S + 1/Y_{fs}$. With a minimum Y_{fs} of 5000 μmhos and a maximum Y_{fs} of 10,000 μmhos, the $1/Y_{fs}$ factors are 200 and 100, respectively. Thus, maximum input impedance is $2200 + 200 = 2400$, with minimum input impedance of $2200 + 100 = 2300$.

coupling capacitors — The value of C_1 forms a high-pass filter with R_{S2}. The low-frequency limit of 30 Hz requires a capacitance value of:

$$C_1 \approx \frac{1}{3.2 \times 30 \times 0.0011} \approx 10 \ \mu F$$

This will provide an approximate 1 dB drop at the low-frequency limit of 30 Hz. If a greater drop can be tolerated, the capacitance value of C_1 can be lowered. The value of C_2 is found in the same manner, except the resistance value must be the load resistance.

2–6.7 Basic MOSFET Amplifier Without Fixed Bias

Figure 2–40 is the working schematic of a basic single-stage MOSFET amplifier without fixed bias. Note that the circuit is similar to the basic self-bias circuits described in Sec. 2–4, except that the input and output coupling capacitors C_1 and C_2 are added, as is resistor R_1. The capacitors prevent direct-current flow to and from external circuits. The resistor R_1 provides a path for bias and signal voltages between gate and source.

Input to the amplifier is applied between gate and ground, across R_1. Output is taken across the drain and ground. The input signal adds to, or subtracts from, the bias voltage across R_1. Variations in bias voltage cause corresponding variations in I_D and the voltage drop across R_L. Therefore, the drain voltage (circuit output) follows the input signal waveform, except that the output is inverted in phase.

Variations in I_D also cause variation in voltage drop across R_S, and a change in the gate-source bias relationship. The change in bias that results from the voltage drop across R_S tends to cancel the initial bias change caused by the input signal and serves as a form of negative feedback to increase stability (and limit gain).

The major difference in the circuit of Fig. 2–40 and a MOSFET amplifier with fixed bias is that the amount of I_D at the Q-point is set entirely by the value of R_S. It may not be possible to achieve a desired I_D with a practical value of R_S. Therefore, it may not be possible to operate at some given value of forward transconductance. If this is of less importance than minimizing the number of circuit components (elimination of one resistor), the circuit of Fig. 2–40 can be used in place of the fixed-bias MOSFET amplifier.

2–6.7.1 *Design Considerations*

Design considerations for the amplifier of Fig. 2–40 are essentially the same as for the fixed-bias amplifier, with the following exceptions.

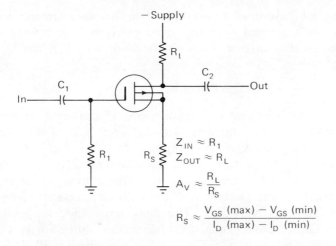

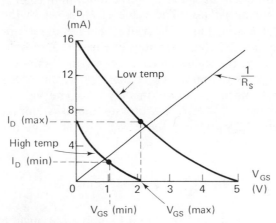

Fig. 2-40 Basic MOSFET amplifier without fixed bias.

If the input and/or output impedances are specified, the resistance values (R_1, R_S, R_L) are set, as shown in Fig. 2–40. However, there are certain limitations for the resistance values imposed by trade-offs (for gain, impedance match, operating point, etc.).

For example, the value of R_S sets the amount of bias, and thus the amount of I_D, while the ratio of R_L/R_S sets the amount of gain. Furthermore, the value of R_L sets the output impedance. If R_S is changed in order to change I_D, both the gain and drain Q-point will change. Also if R_L is changed to match a given impedance, both the gain and Q-point will change.

The input impedance is set by R_1. A change in R_1 will have little effect on gain, operating point, or output impedance. However, R_1 forms a high-pass RC filter with C_1 (if used). A decrease in R_1 requires a corresponding increase in C_1 to accommodate the same low-frequency cut-off point. As a general rule, the value of R_1 is high (in the megohm range), minimizing current drain on the stage ahead of the MOSFET.

2–6.7.2 Design Example

Assume that the circuit under discussion is to be used as a single-stage voltage amplifier without fixed bias. The desired output is 5V (peak-to-peak) with an input of 1V. This requires a gain of at least 5. The MOSFET to be used has a minimum Y_{fs} of 2000 μmhos (a $1/Y_{fs}$ of 500), and V-I characteristics similar to those of Fig. 2–40. The desired I_D at the Q-point is 4 mA with an allowable swing from 2 to 6 mA. The supply is 30V. The input impedance is specified as 3 megohms. The output impedance is unspecified, but must be calculated for reference to other circuits. The low-frequency limit is 30 Hz.

calculate resistance values—The value of R_1 is set by the desired input impedance of 3 megohms. The value of R_S is determined by drawing $1/R_S$ load line on the characteristics of Fig. 2–40, or by the equation, or both. Either way, the value of R_S is set by the limits of V_{GS} and I_D. The value of V_{GSmin} is the point where the I_{Dmin} of 2 mA cross the high-temperature limit curve, or approximately 1V. The value of V_{GSmax} is the point where the I_{Dmax} of 6 mA crosses the low-temperature curve, or approximately 2.3V. Using these values, the first trial value for R_S is:

$$R_S \approx \frac{2.3 - 1}{6 - 2} \approx 325 \text{ ohms}$$

Draw a $1/R_S$ line, starting from the 0 V_G and 0 I_D points, to make sure that it is possible for a straight line to *pass between* the two point, as shown in Fig. 2–40. With the $1/R_S$ line passing through the approximate middle of the two points, any point on the line represents approximately 325 ohms (when the corresponding voltage is divided by the corresponding current).

With the value of R_S set at 325 ohms and a $1/Y_{fs}$ of 500, the value of R_L must be at least 5 times (preferably more) greater to obtain the desired gain of 5. $(325 + 500 = 825;\ 825 \times 5 = 4125$. Use 4.2K ohms.)

With R_L at 4.2K and 4 mA flowing at the Q-point, the drop across R_L is 16.8V, which is approximately one-half the supply of 30V. The drain voltage Q-point of approximately 13.2V will permit a 5V peak-to-peak output swing. Likewise, the six-to-one ratio of 5V output to 30V supply should be sufficient to keep distortion at a minimum. The output impedance of the circuit is approximately 4.2K (the value of R_L).

coupling capacitors—The value of C_1 is:

$$C_1 \approx \frac{1}{3.2 \times 30 \times 3} \approx 0.003 \ to \ 0.004 \ \mu\text{F}$$

sufficient feedback— As a final check of the design values, compare the calculated gain versus the ratio of R_L/R_S. For a highly stable circuit, the gain should be at least 75 percent of the ratio. In this example, the ratio is approximately 13, with the gain slightly over 5; 75 percent of 13 is about 8. Thus the gain is less than 75 percent, and there is insufficient feedback for a truly stable circuit. Of course, this circuit could be adequate for many applications.

There are two possible solutions to the problem of insufficient feedback. First, use a MOSFET with a higher Y_{fs} value (lower $1/Y_{fs}$ factor), making R_S the dominant factor in the gain equation. The other solution is to use a much larger value of R_S. Of course, it may not be possible to use substantially larger R_S and still keep the I_D near the desired limits. From a practical standpoint, the best solution is to use some fixed bias.

2–6.8 Basic MOSFET Amplifier with Zero Bias

Figure 2–41 is the working schematic of a basic single-stage MOSFET amplifier, operating at zero bias and without feedback. Note that the circuit is similar to the basic circuits described in Sec. 2–4, except that the input and output coupling capacitors C_1 and C_2 are added, as is resistor R_1. The capacitors prevent direct-current flow to and from external circuits. The resistor R_1 provides a path for signal voltages between gate and source.

Input to the amplifier is applied across R_1 between gate and ground. Output is taken across the drain and ground. The input signal varies the voltage on the gate. Variations in gate voltage cause corresponding variations in I_D and the voltage drop across R_L. Therefore, the drain voltage (or circuit output) follows the input signal waveform, except that the output is inverted in phase.

Any depletion or depletion-enhancement type MOSFET will have some value of I_D at $0 \ V_{GS}$. If the MOSFET has characteristics similar to those of Fig. 2–41, the I_D will vary between about 1.75 mA and 4 mA, depending on temperature, and from device to device. Thus, with a zero-bias circuit, it is impossible to set the Q-point I_D at any particular value. Likewise, drain voltage Q-point is subject to considerable variation. Since there is no source resistor, there is no negative feedback. Therefore, there is no means to control this variation in I_D. For these reasons, the zero-bias circuit is used where circuit stability is of no particular concern.

The gain of a zero-bias circuit is set by Y_{fs} and the value of R_L, as shown by the equation of Fig. 2–41.

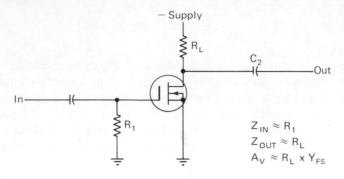

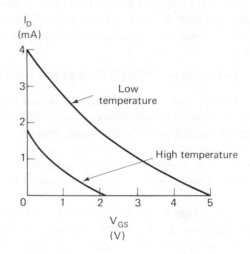

Fig. 2–41 Basic MOSFET amplifier with zero bias.

2–6.8.1 *Design Considerations*

Design considerations for the amplifier of Fig. 2–41 are somewhat similar to the fixed bias amplifier, but with the following exceptions.

If the input and/or output impedances are specified, the resistance values of R_1 and R_L are set as shown. However, there are certain limitations for the resistance values imposed by trade-offs (for gain, impedance match, operating point, etc.).

For example, the value of R_L sets the amount of gain (with a stable Y_{fs}) and the drain voltage operating point (with a stable I_D). Thus, a change in Y_{fs} (usually accompanied by a change in I_D) causes a change in gain (and probably a shift in Q-point voltage). At best, the zero-bias circuit is unstable, even though the input and output impedances remain fairly constant. In designing the zero-bias circuit, both the minimum and maximum values of I_D must be considered, as well as the minimum and maximum Y_{fs}. If the final circuit meets the design requirements at both extremes, the trial values should be satisfactory.

2–6.8.2 *Design Example*

Assume that the circuit of Fig. 2–41 is to be used as a single-stage voltage amplifier without bias or feedback. The desired output is 5V (peak-to-peak) with an input of 0.5V. This requires a gain of at least 10. The MOSFET to be used has a minimum Y_{fs} of 3000 μmhos (0.003 mho) and a maximum Y_{fs} of 4000 μmhos (0.004 mho). The $V-I$ characteristics are similar to those in the figure. The I_D varies between 1.75 mA and 4 mA, depending on temperature and device. The supply is 30V. The input impedance is specified as 3 megohms. The output impedance is unspecified but must be calculated for reference to other circuits. The low-frequency limit is 30 Hz.

operating point—With a supply of 30V, the operating point should be about one-half, or 15V. Since it is impossible to determine the exact value of I_D at any given time, a compromise must be reached. Start by assuming that the voltage drop across R_L must be greater than the minimum peak-to-peak output voltage. With a minimum output of 5V, assume that the minimum drop across R_L must be 7V.

With a minimum I_D of 1.75 mA and a drop of 7V, the value of R_L is 4K. With R_L as 4K and a maximum I_D of 4 mA, the drop across R_L is 16V. Either 7V or 16V operating points are adequate for the 5V output swing.

gain—With R_L at 4K and a minimum Y_{fs} of 0.003, the gain is 12 (4000 × 0.003 = 12). With R_L at 4K and a maximum Y_{fs} of 0.004, the gain is 16. Either gain is sufficient to raise 0.5V to a level greater than 5V. Thus, the basic gain requirements are met. However, assuming that the gain is 16 and the input is 0.5V, the output will be 8V (peak-to-peak). This may or may not cause distortion (or clipping) if the drain is at the 7V Q-point. In practice, an 8V peak-to-peak signal output should vary about 4V on either side of the operating point, or from 3 to 11V around a 7V operating point. The same output signal will vary from about 12 to 20V around the 16V operating point.

input impedance and coupling capacitors—The input impedance of 3 megohms is set by the value of R_1. The value of C_1 is:

$$C_1 \approx \frac{1}{3.2 \times 30 \times 3} \approx 0.003 \text{ to } 0.004 \ \mu\text{F}$$

2–6.9 Multistage MOSFET Amplifiers

In theory, any number of MOSFET amplifier stages can be connected in cascade (output of one amplifier to input of next amplifier) to increase voltage gain. In practice, the number of stages is usually limited to three. The overall gain of the amplifier is the cumulative gain of each stage, multiplied by the gain of the adjacent stage. For example, if each stage of a three-stage amplifier has a gain of 10, the overall gain is 1000 ($10 \times 10 \times 10$). Since it is possible to design a fairly stable single MOSFET stage with a gain of 15 to 20, a three-stage MOSFET amplifier could provide gains in the 3000 to 8000 range. Generally, this is more than enough gain for most practical applications.

2–6.9.1 *Design Considerations*

Any of the single-stage MOSFET amplifiers described in previous sections of this chapter could be connected together to form a two-stage or three-stage voltage amplifier. For example, the basic stage of Sec. 2–6.2 (with a gain of 8) can be connected to two like stages and is cascade. The result is a highly temperature-stable voltage amplifier with a minimum gain of 512 ($8 \times 8 \times 8$). Since each stage has its own feedback, the gain is precisely controlled and very stable.

It is also possible to mix stages to achieve some given design goal. For example, a three-stage amplifier can be designed by using the amplifier of Sec. 2–6.2 (gain of 8) for one stage, and the amplifier of Sec. 2–6.3 (gain of 100) for the remaining two stages. This results in an overall gain of 80,0000. Of course, with bypassed source resistors, the gain is dependent on Y_{fs} and is therefore unpredictable. However, once the gain is established for a given amplifier, the gain should remain fairly stable.

Since design of a multistage MOSFET amplifier is essentially the same as design of individual stages, no specific design example is given. In practical terms, design each stage as described in previous sections of this chapter, then connect the stages together. However, a few precautions must be considered.

feedback—When each stage of a multistage amplifier has its own feedback (local or stage feedback), the most precise control of gain is obtained. However, such feedback is often unnecessary. Instead, overall feedback (or loop feedback) can be used, where a part of the output from one stage is fed back to the input of a previous stage. Usually, such feedback is through resistance (to set the amount of feedback), and the feedback is from the final stage to the first stage. However, it is possible to use feedback from one stage to the next (second stage to first stage, third stage to second stage, etc.).

Be careful to watch for *phase inversion* when using loop or overall feedback. In a MOSFET common-source amplifier, the phase is inverted from input (gate) to output (drain). If feedback is between two stages, the phase is *inverted twice,* resulting in *positive feedback.* This usually produces oscillation. In any event, positive feedback will not stabilize gain.

If feedback between two stages must be used, connect the output (drain) of the second stage back to the source terminal of the first stage, producing the desired *negative feedback.* For example, if the gate (input) of the first stage is swinging positive, the drain of that stage will swing negative, as will the gate of the second stage. The drain of the second stage will swing positive, and this positive swing can be fed back to the source of the first stage. A positive input at the source of a common-source amplifier has the same effect as a negative at the gate. Thus, negative feedback is obtained.

distortion and clipping—As is the case with any high-gain amplifier, be careful not to overdrive the circuit. If the maximum input signal is known, check this value against the overall gain and the maximum allowable output signal swing. For example, assume an overall gain of 1000 and a supply voltage of 20V. With the final stage receiving a 20V supply, the drain voltage Q-point is typically 10V. Theoretically, this will allow a 20V (peak-to-peak) output swing (from 0 to 20V), but, in practice, a swing from about 1 to 19V is more realistic. Either way, a 20 mV (P-P) input signal, multiplied by a gain of 1000, will drive the final output to its limits and possibly into distortion or clipping.

This problem brings up the trade-off between controlled-gain stages (Sec. 2–6.2) and uncontrolled-gain stages (Sec. 2–6.3). If gain is most important, use the uncontrolled-gain stages (or include at least one such stage in the amplifier). If a given output for a given input is the goal, use all controlled-gain stages.

low-frequency cutoff—Since the gate of a MOSFET acts essentially as a capacitor, rather than a diode junction, no coupling capacitor is needed between stages. In theory, this means that there is no low-frequency cutoff problem for ac signals. In practical design, the input capacitance can form an RC filter with the source resistance and can result in some low-frequency attenuation.

The effects of these RC filters are cumulative. For example, if each filter causes a 1-dB drop at some given cutoff frequency, and there are three filters (one at the input and two between MOSFET stages), the result is a 3-dB drop at that frequency in the final output.

A *very approximate* guideline for finding the low frequency at which the gain will drop by 1 dB, for a three-stage MOSFET amplifier follows:

$$\text{frequency} \approx \frac{1}{CR}$$

where frequency is in Hz, C is input capacitance (C_{iss}) in microfarads, and R is source resistance in megohms.

For example, assume that all three MOSFETs in a three-stage amplifier have a C_{iss} of 10 pF, and the source resistance R_S is 1K in all stages. The low frequency at which 1-dB drop occurs is:

$$\text{frequency} \approx \frac{1}{10^{-6} \times 1000} \approx \frac{1}{10^{-3}} \approx 100 \text{ Hz}$$

operating point—Figure 2–42 is the working schematic of a three-stage MOSFET amplifier. Note that all three MOSFETs are of the same type, all three drain resistors (R_1, R_2, R_3) are the same value (to simplify design), and no source resistors are used. Such an amplifier will provide

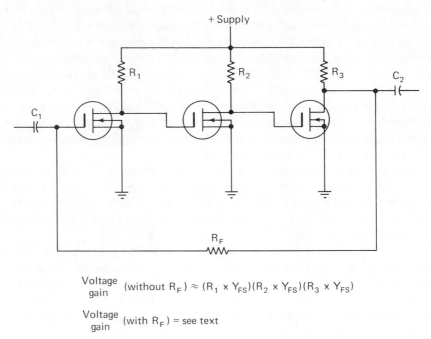

$$\frac{\text{Voltage}}{\text{gain}} \text{ (without } R_F) \approx (R_1 \times Y_{FS})(R_2 \times Y_{FS})(R_3 \times Y_{FS})$$

$$\frac{\text{Voltage}}{\text{gain}} \text{ (with } R_F) = \text{see text}$$

Fig. 2–42 Basic MOSFET multistage amplifier.

maximum gain, but minimum stability. At first glance, it may appear that all three stages are operating at zero bias. However, when I_D flows, there is some drop across the corresponding drain resistor, producing some voltage at the drain of the stage, and an identical voltage at the gate of the next stage. The gate of the first stage is at the same voltage as the drain of the last stage, because of feedback resistor R_F. There is no current drain through R_F, with the possible exception of reverse gate current (which can be ignored).

To find a suitable operating point for the amplifier, it is necessary to trade off between desired output voltage swing, device characteristics, and supply voltage. For example, assume that an output swing of 7V peak-to-peak is desired, and the supply voltage is 24V. Further assume that the I_D is about 0.5 mA when V_{GS} is 7V. A suitable operating point would be 7V, accommodating the 7V output swing without distortion. (The swing will be from about 3.5 to 10.5V, centered on the 7V Q-point.) This requires an 18V drop from the 24V supply. With 0.55 mA, I_D, and an 18V drop, the values of R_1, R_2 and R_3 are about 33,000 ohms.

gain—The overall voltage gain is dependent upon the relationship of the gain without feedback, and the feedback resistance R_F. Gain without feedback is determined by Y_{fs} and the values of R_1, R_2 and R_3. For example, assume a Y_{fs} of 1000 μmhos (0.001 mho). The gain of each stage is 33; $33,000 \times 0.001 = 33$. With each stage at a gain of 33, the overall gain (without feedback) is about 36,000.

To find the value of R_F, divide the gain without feedback by the desired gain. Multiply the product by 100. Then multiply the resultant product by the value of R_1. For example, assume a desired gain of 3000. (The gain without feedback is 36,000): $36,000/3000 = 12$; $12 \times 100 = 1200$; $1200 \times 33,000 = 39.6M$ (use the nearest standard to 40M).

input impedance—The input impedance is dependent on the relationship of gain and feedback resistance. The approximate impedance is: R_F/gain.

Since gain is dependent on Y_{fs}, input impedance is subject to variation with temperature and from device to device.

2–6.10 Direct-Current MOSFET Amplifier

The circuit of Fig. 2–42 requires one coupling capacitor at the input to isolate the input gate from any direct current voltage that may appear at the input generator or other device. This makes this circuit unsuitable for use as a direct-current amplifier (also known as a direct-coupled amplifier).

The circuit can be converted to a direct-coupled amplifier when the coupling capacitor is replaced by a series resistor R_{in}, as shown in Fig. 2–43. The considerations concerning operating point are the same for both circuits (Figs. 2–42 and 2–43). However, the series resistance R_{in} must be terminated at a dc level *equivalent to the operating point*. For example, if the operating point is 7V, point A must be at 7V. If point A is at some other dc level, the operating point of the amplifier will be shifted.

The relationship of input impedance, R_F and gain still holds. However, the input impedance is approximately equal to R_{in}. Therefore, gain is approximately equal to the ratio of R_F/R_{in}. This makes it possible to control gain by setting the R_F/R_{in} ratio. Of course, the gain cannot exceed the

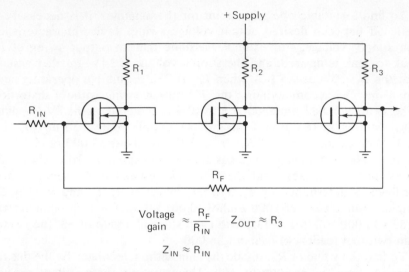

Fig. 2–43 Basic MOSFET direct-current (direct-coupled) amplifier.

gain-without-feedback (open loop) factor, no matter what the ratio of R_F/R_{in} (closed loop). Generally, the greater the ratio of open loop gain to closed loop gain, the greater the circuit stability.

For example, assume that the desired gain is 5000, R_F is 40M, and the open-loop gain is 36,000. Since 5000 is considerably less than 36,000, the circuit is easily capable of producing the desired gain with feedback. To find the value of R_{in}:

$$R_{in} \approx \frac{40 \times 10^6}{5 \times 10^3} \approx 8 \times 10^3 \qquad (8000 \text{ ohms})$$

2–6.11 Amplification from Grounded Sources

The circuit in Fig. 2–43 requires that the signal source be at a dc level equal to the operating point. In many cases, it is necessary to amplify direct-current signals at the zero or ground level. This can be done with a depletion-type MOSFET at the input, as shown in Fig. 2–44.

Feedback is introduced by connecting the sources of both Q_1 and Q_3 to a common-source resistor R_2. The source of Q_2 is not provided with a source resistor, but there is some bias on Q_2 produced by the I_D drop across R_3. The input impedance of the Fig. 2–44 circuit is set by the value of R_1 and by the gate-drain capacitance of Q_1.

Gain can be sacrificed for stability by increasing the value of R_2. With the values shown and typical MOSFETs, the gain should be in the order of 3000 to 5000. The bias and operating point for Q_2 and Q_3 is set by R_3,

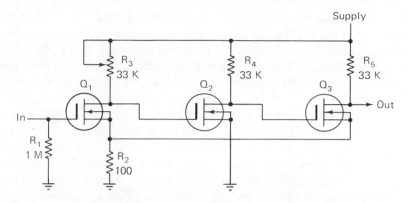

Fig. 2–44 MOSFET circuits for amplification of direct-current signals at zero or ground level.

shown as 33K. In practice, the value of R_3 is approximated by calculating, and then adjusted for a desired operating point at the output (drain) of Q_3.

2–6.12 Hybrid Direct-Coupled Amplifiers

In certain applications, MOSFET stages and two-junction transistor stages can be combined to form hybrid amplifiers. The classic example is where a single MOSFET stage is used at the input, followed by two two-junction amplifier stages. Such an arrangement takes advantage of both the MOSFET and two-junction transistor characteristics.

A MOSFET is essentially a voltage-operated device, permitting large voltage swings with low currents. This makes it possible to use high resistance values (resulting in high impedance) at the input and between stages. In turn, these high resistance values permit the use of low-value coupling capacitors and eliminate the need for bulky expensive electrolytic capacitors. However, MOSFETs have the characteristic of operating at low currents, and so are considered as low-power devices.

Two-junction transistors are essentially current-operated devices, permitting large currents at about the same voltage levels as the MOSFET. Thus, with equal supply voltage and signal voltage swings, the two-junction transistor can supply much more current gain (and power gain) than the MOSFET. Since currents are high, the impedances (input, interstage, output) must be low in two-junction transistor amplifiers. This requires large-value coupling capacitors if low frequencies are involved. The low impedances also place a considerable load on devices feeding the amplifier, particularly if the devices are high impedance. On the other hand, a low output impedance is often a desirable characteristic for an amplifier.

When a MOSFET is used as the input stage, the amplifier input impedance is high, placing a small load on the signal source, and allowing the use of a low-value input coupling capacitor (if required). If the MOSFET is operated at the OTC point (Sec. 2–2), the amplifier input is temperature stable. (Generally, the input stage is the most critical in regards to temperature stability.) When two-junction transistors are used as the output stages, the output impedance is low, and current gain (as well as power gain) is high.

Hybrid amplifiers can be direct-coupled or capacitor-coupled, depending on requirements. The direct-coupled configuration offers the best low-frequency response, permits direct-current amplification, and generally is simpler (uses less components). The capacitor-coupled hybrid amplifier permits a more stable design and eliminates the voltage regulation problem common to all direct-coupled amplifiers. (That is, a direct-coupled amplifier cannot distinguish between changes in signal level and changes in power-supply level.)

The MOSFET can be combined with any of the classic two-stage, two-junction transistor amplifier combinations. The two most common combinations are the Darlington pair (for no voltage gain, but high current gain and low output impedance), and the *NPN-PNP* complementary-amplifier pair (for both voltage gain and current gain).

2–6.12.1 *Hybrid Amplifier with MOSFET Input and Two-Junction Output*

Figure 2–45 is the working schematic of a direct-coupled amplifier using a MOSFET input stage and a two-junction transistor pair as the output. Note that local feedback is used in the MOSFET stage (provided by source resistor R_S), as well as *overall feedback* (provided by resistance R_4).

The design considerations for the MOSFET portion of the circuit are essentially the same as described in previous sections of this chapter, with certain exceptions. Input impedance is set by the value of R_2, as usual. Output impedance is set by the combination of R_4 and R_S. However, since R_S is quite small in comparison to R_4, the output impedance is essentially equal to R_4.

The gain of the MOSFET stage is set by the ratio of R_L to R_S, plus the $1/Y_{fs}$ factor. However, since R_S is quite small, the MOSFET gain is set primarily by the ratio of R_L to $1/Y_{fs}$. The gain of the two-junction transistor pair is set by the beta of the two transistors and by the feedback. Thus, the gain can only be estimated.

Note that the drop across R_3 is the normal base-emitter drop of a transistor (about 0.5 to 0.7V for silicon, and 0.2 to 0.3 for germanium). The drop across R_L is twice this value (about 1 to 1.5V for silicon, and 0.4 to

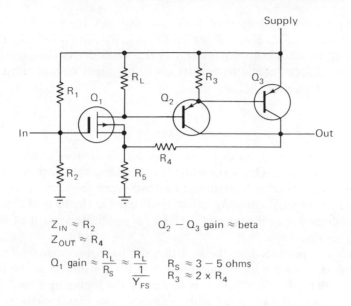

$$Z_{IN} \approx R_2 \qquad\qquad Q_2 - Q_3 \text{ gain} \approx \text{beta}$$
$$Z_{OUT} \approx R_4$$
$$Q_1 \text{ gain} \approx \frac{R_L}{R_S} \approx \frac{R_L}{\frac{1}{Y_{FS}}} \qquad R_S \approx 3 - 5 \text{ ohms}$$
$$R_3 \approx 2 \times R_4$$

Fig. 2–45 Hybrid amplifier with MOSFET input and two-junction transistor output.

0.6V for germanium). For a typical silicon transistor, the base of Q_2 and the drain of Q_1 operate at about 1V removed from the supply. In a practical experimental circuit, R_L must be adjusted to give the correct bias for Q_2 (and operating point for Q_1). The same is true for R_3. However, as a first trial value, R_3 should be about twice the value of R_4.

Design starts with a selection of I_D for the MOSFET. If maximum temperature stability is desired, use the OTC level of I_D. This usually requires a fixed bias, as previously described. If temperature stability is not critical, the MOSFET can be operated at zero bias by omitting R_1. There is some voltage developed across R_S. However, since R_S is small, the V_{GS} is essentially zero, and the I_D is set by the zero V_{GS} characteristic of the MOSFET. With the value of I_D set, select a value of R_L that produces approximately 1 to 1.5V drop to bias Q_2.

The input impedance is set by R_2, with the output impedance set by R_4. The value of R_3 is approximately twice that of R_4. The value of R_S is less than 10 ohms, typically in the order of 3 to 5 ohms.

As a brief design example, assume that the circuit of Fig. 2–45 is to provide an input impedance of 1 megohm, and output impedance of 500 ohms, and maximum gain. Temperature stability is not critical. Under these conditions, the values of R_2 and R_4 are set at 1M and 500 ohms (or the nearest standards). The value of R_S is 5 ohms, but the voltage drop across R_S can be ignored. Assume that I_D is 0.2 mA under these conditions. With a required drop of 1.5V, and 0.2 mA I_D, the value of R_L is approximately 7.5K. Since R_4 is 500 ohms, R_3 should be 1K. The key component in setting up this circuit is R_L. With the circuit operating in experimental form, adjust R_L for the desired Q-point voltage at the output (collectors of Q_2 and Q_3).

2–6.12.2 *Non-Blocking Direct-Coupled Amplifier*

Generally, a direct-coupled amplifier does not require any coupling capacitors. One exception is a coupling capacitor at the input (to isolate the amplifier from direct current) when the signal is composed of dc and ac. When a coupling capacitor is used at the base of a two-junction transistor (or at the gate of a JFET) a condition known as *blocking* possibly can occur.

Blocking is produced by the fact that the base junction of a two-junction transistor (or the gate junction of a JFET) is similar to the junction of a diode. That is, the diode acts to rectify the incoming signal. If a capacitor is connected in series with the diode (gate junction), large signals can charge the capacitor. On one half-cycle, the diode is forward biased and charges rapidly; on the opposite half-cycle, the diode is reverse-biased and discharges slowly. If the signal and charge are large enough, the amplifier can be biased at or beyond cutoff, until the capacitor discharges. Thus, the amplifier can be blocked to incoming signals for a period of time.

When a MOSFET is used at the input of an amplifier, the blocking problem is eliminated since the MOSFET gate is similar to a capacitor. That is, the MOSFET gate does not act like a diode, and there is no rectification of the signal. In effect, the gate charges and discharges at the same rate as the coupling capacitor.

Figure 2–46 shows a nonblocking amplifier with a MOSFET input driving a two-junction transistor pair. The input impedance is set by R_{IN}, with the output impedance set by R_7. With the resistance ratios shown by the equation of Fig. 2–46, the voltage gain is approximately 10 when capacitor C_2 is *out of the circuit*, and about 1000 when C_2 is in the circuit. Keep in mind that feedback is reduced or removed (and gain is increased) when C_2 is in the circuit, since capacitor C_2 functions to bypass feedback signals to ground. With C_2 removed, the full feedback is applied, and gain is minimum (stability is maximum).

If it is desired to operate the amplifier at some gain level between 10 and 1000, use a resistance in series with C_2 (shown in phantom as R_8).

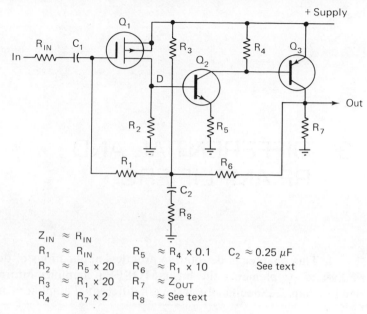

$$Z_{IN} \approx R_{IN}$$

$R_1 \approx R_{IN}$	$R_5 \approx R_4 \times 0.1$	$C_2 \approx 0.25\ \mu F$
$R_2 \approx R_5 \times 20$	$R_6 \approx R_1 \times 10$	See text
$R_3 \approx R_1 \times 20$	$R_7 \approx Z_{OUT}$	
$R_4 \approx R_7 \times 2$	$R_8 \approx$ See text	

Fig. 2–46 Nonblocking direct-coupled amplifier with MOSFET input.

3. DIFFERENTIAL AND RF AMPLIFIERS

This chapter is devoted to practical applications of discrete MOSFETs and supplements the basic MOSFET amplifier information described in Chap. 2. Specifically, it covers differential amplifiers and RF amplifiers.

It is necessary to understand the basics of such circuits to make full use of MOSFET advantages, and to avoid MOSFET limitations. For this reason, the basic principles of the circuits are discussed briefly in the following sections. Such discussions are slanted as to how MOSFETs fit into the particular circuit.

3–1. DIFFERENTIAL AMPLIFIERS

In a theoretical differential amplifier, no output is produced when the signals at the inputs are identical. That is, an output is produced *only when there is a difference* in signals at the input. A differential amplifier has two inputs and two outputs, although only one output may be used in many applications.

One of the main uses for differential amplifiers is as the input stage for an operational amplifier (Op-amp), particularly integrated-circuit op-amps. Another use for differential amplifiers in laboratory work is as an amplifier for meters, oscilloscopes, recorders, and the like. Such instruments are operated in areas where many signals may be radiated (power-line radiation, stray signals from generators, etc.). Test leads connected to the input terminals would pick up these radiated signals, even if the leads were shielded. If a single-ended input (basic amplifier) were used, the undesired signals would be picked up and amplified along with the desired signal input. If the amplifier has a differential input, both leads will pick up the same radiated signal at the same time. Since there is no difference between the radiated signals at the two inputs, there is no amplification of the undesired inputs.

Signals common to both inputs (such as radiated signals) are known as *common-mode signals*. The ability of a differential amplifier to prevent conversion of a common-mode signal into a difference signal (which produces an output) is expressed by its *common-mode rejection ratio* (CMR or CMRR). Common-mode terms are discussed in later parts of this section.

3–1.1 Basic Differential Amplifier Theory

Figure 3–1 is the schematic of a basic differential amplifier.

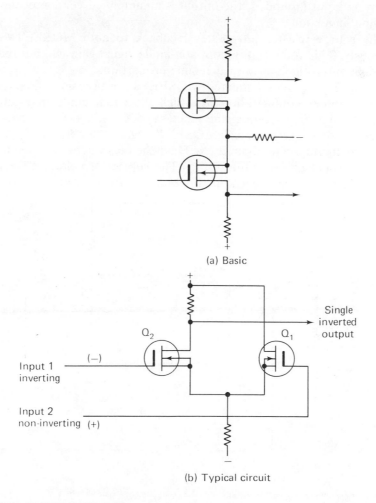

(a) Basic

(b) Typical circuit

Fig. 3–1 Basic differential amplifier circuits.

The circuit responds differently to common-mode signals than it does to single-ended signals.

A common-mode signal (like power-line pickup) drives both gates in phase with equal-amplitude ac voltages, and the circuit behaves as though the MOSFETs are in parallel to cancel the output. In effect, one MOS-FET cancels the other.

Normal signals are applied to either of the gates (Q_1 or Q_2). The *inverting input* is applied to the gate of Q_2, and the noninverting input is applied to the gate of Q_1. With a signal applied only to the inverting input, and the noninverting input grounded, the output is an amplified and inverted version of the input. For example, if the input is a positive pulse, the output is a negative pulse. If the noninverting input is used with the inverting input grounded, the output is an amplified version of the input (without inversion).

The source resistor introduces feedback to both MOSFETs simultaneously. This reduces the common-mode signal gain without reducing the differential signal gain in the same proportion.

In Fig. 3–2, triode-connected MOSFETs are used in a simple differential amplifier configuration in which the triode-connected gates are biased from a single source (the junction of R_1 *and* R_2). This arrangement is possible because a typical MOSFET has a gate current (I_{GSS}) in the triode configuration of about 2nA. Thus, the bias can be supplied through R_3 with a negligible *voltage offset*. (The subject of voltage offset is discussed further in later parts of this section.) Resistor R_5 is used to null

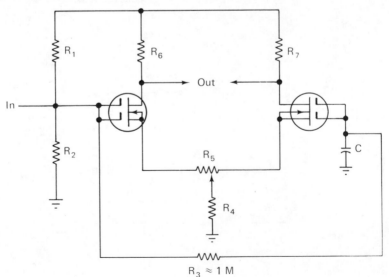

Fig. 3–2 Simple differential amplifier circuit using triode-connected dual-gate MOSFETs.

out the effects of slight differences in device characteristics so that the voltage offset is zero (with zero input voltage).

The circuit in Fig. 3–3 shows another differential amplifier configuration, in which the voltage offset can be set to zero by means of appropriate potentials supplied to the # 2 gates, with adjustment provided by R_6.

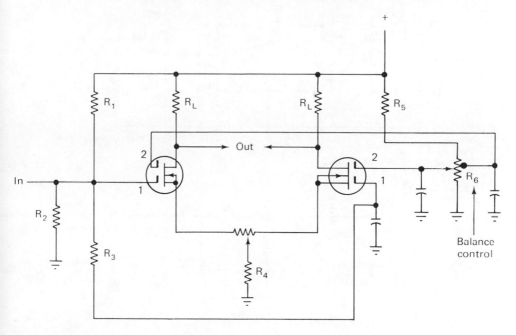

Fig. 3–3 Differential amplifier using dual-gate MOSFETs with gate 2 as balance control input.

Figure 3–4 is the schematic of a differential amplifier typical of those found in laboratory instruments. The circuit is basically a single-stage differential amplifier (Q_2 and Q_4) with MOSFET input source-followers (Q_1 and Q_5), and *constant-current* source Q_3 in the emitter-coupled leg. Note that this circuit is a hybrid arrangement, such as that described in Chap. 2, since both MOSFET and two-junction devices are used. Also note that the single emitter resistor of the Fig. 3–1 circuit is replaced by Q_3 and its associated components in the circuit of Fig. 3–4.

The use of a transistor such as Q_3 is typical for many differential amplifiers, particularly those found in operational amplifiers (as the first stage). The circuit of transistor Q_3 is known as a *temperature-compensated constant-current source*. All current for the differential amplifier is fed through Q_3 (an NPN) connected between the emitters of the differential amplifier and V_{EE} (the negative power supply). If there is an increase in current, a larger voltage is developed across the current-source Q_3 emitter resistor. This larger voltage acts to reverse bias the base emit-

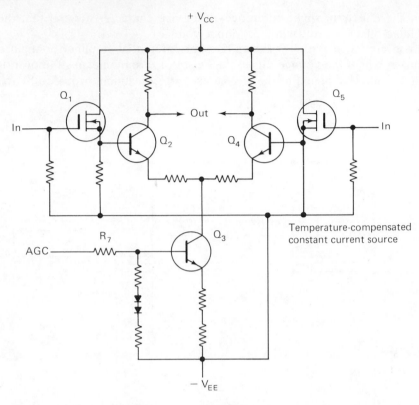

Fig. 3–4 Typical hybrid differential amplifier with constant-current source.

ter, therefore reducing current through Q_3. Since all current for the differential amplifier is passed through Q_3, current to the amplifier is also reduced. If there is a decrease in current, the opposite occurs, and the amplifier current increases. Thus, the differential amplifier is maintained at a constant current level.

Transistor Q_3 is also *temperature compensated* by diodes connected in the base-emitter bias network. These diodes have the same (approximate) temperature characteristics as the base-emitter junction, and offset any change in Q_3 base-emitter current flow that results from temperature change.

3–1.2 MOSFET Versus Two-Junction Transistors

Figure 3–5 shows the basic circuits involved for both two-junction and MOSFET differential amplifiers. With either circuit, *floating signals* not referenced to ground can be amplified and, since large values of common-mode rejection can be achieved, small differential signals can

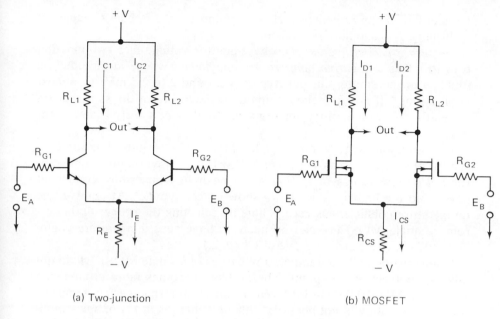

(a) Two-junction   (b) MOSFET

Fig. 3-5 Comparison of MOSFET and two-junction differential amplifier circuits.

be discriminated from large common-mode signals on which the differential signals may be riding.

Also, because of the matching and temperature tracking that can be achieved between the devices in the differential amplifier (particularly in ICs and in devices on the same chip), the circuit can have very good dc stability over a very wide temperature range.

In certain applications, the MOSFET has advantages over the two-junction transistor. The main advantage of the MOSFET in differential amplifier applications is the high-input impedance. As a comparison, whereas differential amplifier input impedance of 100K to 1M can be obtained with two-junction transistors, input impedances ranging upwards of 10^9 are easily within reach of MOSFET amplifiers.

matching requirements—In designing a differential amplifier, either two-junction or MOSFET, some degree of matching is required between certain characteristics of the two transistors.

In two-junction amplifiers, the dc current gain (h_{FE}) and the base-emitter voltage (V_{BE}) of the two transistors should be matched. The collector-base leakage currents (I_{CBO}) should also be nearly equal, although this parameter is generally not too critical, especially in silicon transistors where I_{CBO} is very small.

The MOSFET differential amplifier requires a match of the forward transconductance (Y_{fs}) and the gate-source voltage V_{GS} of the two devices.

In addition, if the impedance of the driving source is high (1 megohm or higher), I_{GSS} should be matched.

temperature coefficients—The base-emitter voltage of a two-junction transistor has a *negative temperature coefficient*, whose magnitude is a function of the emitter current (typically around 2 to 2.5 mV/°C in most applications). If two matched chips are selected from the same silicon wafer, the two base-emitter junctions will track over a fairly wide current range.

The situation is quite different for MOSFETs. Figure 3–6 shows a typical plot of I_D versus V_{GS} for a MOSFET at three different temperatures. For low values of I_D, V_{GS} has a positive temperature coefficient (TC), and the higher values of I_D show a negative TC. At point 0, the temperature coefficient is zero. Thus, by selecting the proper drain current, a single MOSFET can be made to have zero temperature coefficient ($0TC$). (Refer to Sec. 2–2 of Chap. 2.)

Unfortunately, all the temperature curves of a single MOSFET do not always intersect at one point. The problem becomes more pronounced when two MOSFETs are involved. In practical terms, perfect temperature compensation is not possible, although changes in V_{GS} of a few millivolts can be achieved. If two well-matched MOSFETs are used in a differential-amplifier configuration, as shown in Fig. 3–5(b), these minor changes can be balanced out quite effectively.

Finding the Zero Drift Point for Differential-Amplifier

MOSFETs—Theoretically, if the V_{GS} versus I_D characteristics of the two devices in a MOSFET differential amplifier are well-matched, the differential amplifier can be operated at any current level (within the limits of the $V_{GS} = I_D$ curve) with little or no drift. Unfortunately, the degree of matching required is generally not practical from an economical point of view.

In order to minimize drift in a MOSFET differential amplifier, it is very desirable to operate the MOSFETs at or near their zero drift points. Matching can be fairly easily obtained at this one point, and because the MOSFETs are operated with essentially $0TC$, the effects of any mismatch are greatly minimized.

At the zero drift point, V_{GS} and I_D are given by:

$$V_{GS(Z)} = V_P - 0.63 \tag{3-1}$$

$$I_{D(Z)} = I_{DSS}\left(\frac{0.63}{V_P}\right)^2 \tag{3-2}$$

where V_P is the pinch-off voltage and I_{DSS} is the drain current at zero gate-source voltage ($V_{GS} = 0$). The pinch-off voltage is defined as that point where a further increase in V_{DS} causes little change in I_D.

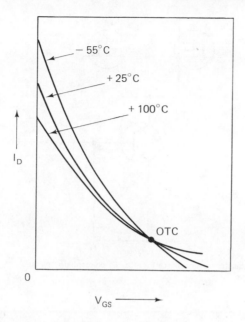

Fig. 3–6 Drain current versus gate-source voltage for typical MOSFET showing OTC point.

varying I_{CS} to achieve minimum drift—Equations 3–1 and 3–2 are based on some assumptions, and as such are subject to some degree of error. The major assumption is that the voltage drops across the input resistances R_G due to I_{GSS} either completely cancel each other, or are so small as to be negligible. Usually, the latter is the case. However, even when the possible errors are considered, the equations are useful in arriving at values of V_{GS} and I_D that are reasonably close to the zero drift point.

In practical design, additional testing (over a given temperature range) becomes necessary to establish the minimum drift point. By varying the current I_{CS} [shown in Fig. 3–5(b) as the sum of the two drain currents], drift on the order of a few hundred microvolts per degree Celsius (referred to as input) can be obtained.

In general, the technique of setting I_{CS} to some given value for drift compensation is not satisfactory. The compensation is usually not as good as desired. More important, the resistor R_{CS} of Fig. 3–5(b) is usually replaced by a current source transistor, as discussed previously. A typical circuit is shown in Fig. 3–7. The zener diode voltage is chosen so that it compensates for variations in the current-source transistor base-emitter junction over a given temperature range. The current-source transistor provides higher common-mode rejection and greatly reduces the effects of power-supply variations on drift. It is very difficult to set the value of I_{CS} if a current-source transistor is used. A description of a better method of drift and offset compensation follows.

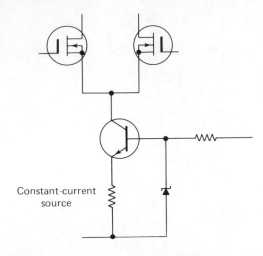

Fig. 3–7 Two-junction transistor current source for MOSFET differential amplifier.

3–1.3 Drift and Offset Compensation Circuit

Figure 3–8 shows an expanded view of the region around the $0TC$ point of the typical MOSFET differential amplifier. For clarity, only two temperatures are shown, and the differences between the two devices have been greatly exaggerated. The $0TC$ points are at A and B. Assuming negligible drop across the input resistances, R_{g1} and R_{g2}, the gate-to-source voltage of the two MOSFETs is equal.

At temperature T_1, the value of V_{GS} (V_{GS1}) is established by the circuit so that the sum of the two drain currents I_{D1} and I_{D2} is equal to I_{CS}. At temperature T_2, V_{GS} shifts to a new value (V_{GS2}) so that, again, I_{D1} plus I_{D2} equals I_{CS}. This change in I_D results in a drift at the differential-amplifier output, since I_{D1} has decreased and I_{D2} has increased.

The *condition for zero drift* at the differential amplifier output is:

$$\Delta I_{D1} R_L = \Delta I_{D2} R_L \tag{3–3}$$

since

$$I_{D1} + I_{D2} = I_{CS} \tag{3–4}$$

and

$$\Delta I_{CS} = 0 \quad \text{(constant current)} \tag{3–5}$$

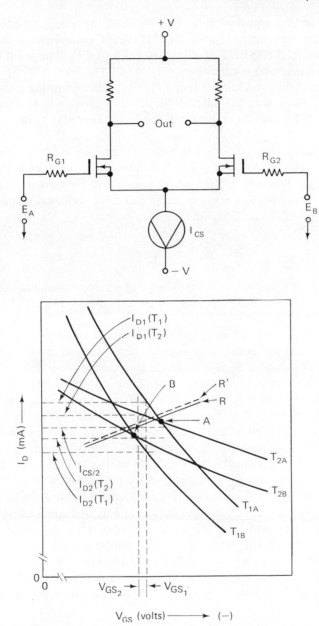

Fig. 3–8 Temperature compensation in MOSFET differential amplifier. (Courtesy Motorola)

Then, for zero drift

$$\Delta I_{D1} = \Delta I_{D2} = 0 \qquad\qquad \textbf{(3–6)}$$

In order to compensate the circuit for drift, operating points must be found so that the drift in V_{GS} for one MOSFET compensates the drift in V_{GS} of the other MOSFET at constant values of I_D. The solid line R of Fig. 3–8 defines such a condition.

As the temperature changes from T_1 to T_2, the operating point moves from the solid line R to the dashed line R'. The change in V_{GS} is equal for both MOSFETs, and there is no change in I_D. This results in a condition of no drift.

The stable operating points are accomplished by *adding a resistor* in the source of the proper MOSFET. The added resistor goes in the source of the MOSFET in which the current *must be decreased*. In the case of the Fig. 3–8 circuit, the resistor goes in the left side, corresponding to I_{D1}. The value of R can be calculated from

$$R = \frac{V_{GS1} - V_{GS2}}{I_D} \tag{3-7}$$

where V_{GS1} and V_{GS2} are the value of gate-to-source voltage at the desired operating points, and I_D is the current through the MOSFET that has the ·added R.

practical drift-offset compensation circuit—In most practical applications there is already some resistance in the source, as shown in Fig. 3–9. The source resistances R_S are included to stabilize the amplifier gain. To produce drift-offset compensation, one of the source resistors is shunted. The shunted resistor is the one in the *side opposite* that in which a compensating resistor is added. For example, in the Fig. 3–8 circuit the source resistor is to be added at the left side. In the Fig. 3–9 circuit the shunt resistor is added across R_{S2} (the right side). The difference between the shunted resistor (R_{S2} in this case) and the unshunted resistor (R_{S1}) is the value of the compensating resistor that would be added (Equation 3–7).

Although the value of compensating resistance required can be approximated using Equation 3–7 and the curves of Fig. 3–8, the best results are obtained by a trial and error selection. Drifts of a few microvolts per degree Celsius can be obtained over a wide temperature range.

voltage gain of compensated differential amplifier—Voltage gain of the MOSFET differential amplifier is the same as that for a single common-source stage, if both sides are well matched. The gain is expressed by

$$A_{dd} = \frac{Y_{fs}R_L}{1 + Y_{os}R_L} \tag{3-8}$$

where A_{dd} is the differential-input to differential-output gain, Y_{fs} is the device forward transadmittance, Y_{os} is the device output admittance, and R_L is the load resistor (equal for each side) of the circuit.

eliminating offset voltage in the differential amplifier—After the drift

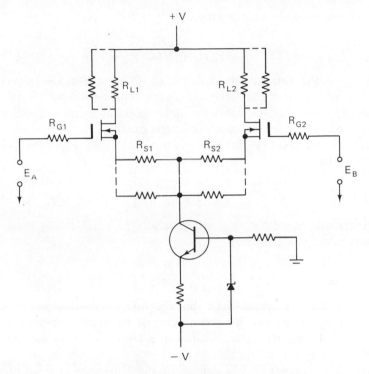

Fig. 3-9 Practical drift-offset compensation for MOSFET differential amplifier. (Courtesy Motorola)

compensation has been accomplished, the voltage offset problem must be considered. Before going into the circuit details, we will define some voltage offset terms.

Input offset voltage (V_{io}) of a differential amplifier is defined as the voltage that must be applied at the input terminals to obtain zero output voltage. Input offset voltage indicates the matching tolerance in differential amplifier stages. A perfectly matched amplifier requires zero input voltage to produce zero output voltage. Input offset voltage affects design in that the input signal must overcome the offset voltage before an output is produced. For example, if V_{io} is 1 mV and a 1-mV signal is applied, there is no output. If the signal is increased to 2 mV, the amplifier will produce only the peaks.

Note that V_{io} is increased by amplifier gain. Thus, what may be a small value of offset at the input can produce a large value of offset at the output. As a guideline, V_{io} is increased by amplifier gain plus unity. For example, if the gain is 100, the effect of input offset voltage is increased by 101.

In most applications, 1 is much greater than $Y_{os}R_L$, so the gain becomes

$$A_{dd} \approx Y_{fs}R_L \tag{3-9}$$

The *common-mode gain* is given by

$$A_{cc} = \frac{Y_{fs}R_L}{1 + Y_{os}R_L + 2R_{CS}(Y_{os} + Y_{fs})} \qquad (3\text{--}10)$$

where A_{cc} is the common-mode-input to differential-mode-output gain, and R_{CS} is the value of the resistance in the common-source circuit of the two MOSFETs. This indicates the desirability of using a constant current source to minimize the common-mode gain.

Since, in most cases, Y_{fs} is much greater than $Y_{os}R_L$, Equation 3–10 can be reduced to

$$A_{cc} \approx \frac{Y_{fs}R_L}{1 + 2R_{CS}Y_{fs}} \qquad (3\text{--}11)$$

Furthermore, since $2R_{CS}Y_{fs}$ will generally be much larger than 1, the common-mode gain becomes

$$A_{cc} \approx \frac{R_L}{2R_{CS}} \qquad (3\text{--}12)$$

Again, it must be emphasized that the foregoing equations for calculating gain are based on the assumption of well-matched devices. The accuracy of the equations is dictated to a large extent by the degree of matching.

Input offset current (I_{io}) is defined as the difference in input bias current into the input of a differential amplifier. It is an indication of the degree of matching of the differential stage. When the impedances feeding the differential stage are high, input offset current can be of greater importance than input offset voltage. If the input bias current is different for each input, the voltage drops across the input resistors (or input impedance) will not be equal. If the resistance is large, there will be a large unbalance in input voltages.

Input bias current (I_i) is sometimes used in differential amplifier specifications. I_i is defined as the average value of the two input bias currents of the input stage. In use, the only real significance of input bias current is that the resultant voltage drop across the input resistors can restrict the input common-mode voltage range at higher impedance levels. Again, the input bias current produces a voltage drop across the input resistors. This voltage drop must be overcome by the input signal.

In a practical circuit, if one source has a different resistance value than the other source, there is a fixed difference in dc voltage at the outputs (drains), as well as a difference in gains. This difference can be offset by shunting one of the load (drain) resistors (R_{L1} or R_{L2}) as shown in Fig. 3–9.

This shunt resistance causes a slight, but usually negligible, variation in gain between the two sides, and offsets the difference produced by the different source resistances. All other factors being equal, the gain of each side is set by the R_L/R_S ratio. If R_{S2} is shunted to provide drift compensation, the value of R_{S2} is reduced and the gain is increased, so that the right side of the differential amplifier produces more gain than the left side. This can be offset by shunting R_{L2} to restore the R_L/R_S ratio, and to make the gain for both sides equal.

3–1.4 Common-Mode Definitions

The terms *common mode* and *common-mode rejection* are used frequently in differential amplifier applications. All manufacturers do not agree on their exact definition.

One manufacturer defines common-mode rejection (CMR, or sometimes listed as CM_{rej}), or the common-mode rejection ratio (CMRR), as the ratio of differential gain (usually large) to common-mode gain (usually a fraction). That is, the amplifier may have a large gain of differential signals (different signals at each input terminal, or one input terminal grounded and the opposite input terminal with a signal), but little gain (or possibly a loss) of common-mode signals (same signal at both terminals).

Another manufacturer defines CMR as the relationship of *change* in output voltage to the *change* in the input common-mode voltage producing it, divided by the open-loop gain (amplifier gain without feedback).

For example, using the latter definition, assume that the common-mode input (applied to both terminals simultaneously) is 1V, the resulting measured output is 1 mV, and the open-loop gain is 100. The CMR is then

$$\frac{\text{(output/input)}}{\text{open-loop gain}} = \text{CMR}$$

$$\frac{(0.001/1)}{100} = 100,000$$

Another method for calculating CMR is to divide the output signal by the open-loop gain to find an *equivalent differential input signal*. Then the common-mode input signal is divided by this equivalent differential input signal. Using the same figures as in the previous CMR calculation,

$$\frac{0.001}{100} = 0.000001, \qquad \text{equivalent differential input signal}$$

$$\frac{1}{0.000001} = 100,000$$

Common-mode rejection is usually specified in decibels (dB). Thus, a

ratio of 100,000 is equal to 100 dB. No matter what basis is used for calculation, CMR is an indication of the *degree of circuit balance* of the
differential stages, since a common-mode input signal should be amplified
identically in both halves of the circuit. A large output for a given common-mode input is an indication of large unbalance or poor CMR. If there
is an unbalance, a common-mode signal becomes a differential signal after
it passes the first stage (or at the output in a single-stage differential amplifier).

As with amplifier gain, CMR usually decreases as frequency increases.
However, as a guideline, the CMR of any differential amplifier should be
at least 20 dB *greater* than the open-loop gain at any frequency (within
the limits of the amplifier).

3–2. RF AMPLIFIERS

MOSFET RF amplifiers can be designed by using two-port
networks similar to those of two-junction transistors. Basically, the method
consists of characterizing the MOSFET as a linear active two-port network (LAN) with admittances (*y* parameters), and using these parameters
to solve exact design equations for stability, gain, and input/output admittances.

It is difficult, at best, to provide a simple, step-by-step procedure for
designing MOSFET RF amplifiers to meet all possible circuit conditions.
In practice, the procedure often results in considerable trial and error.
There are several reasons for this problem of MOSFET RF amplifier
design.

First, not all of the MOSFET characteristics are always available in
datasheet form. For example, input and output admittance may be given
for some low frequency, but not at the desired amplifier operating frequency.

Often, manufacturers do not agree on terminology. A classic example
of this is in *y*-parameters, where one manufacturer uses letter subscripts
(y_{fs}) and another manufacturer uses number subscripts (Y_{21}). Of course,
this can be solved by conversions, described later in this section.

In some cases, manufacturers will give the required information on
datasheets, but not in the required form. For example, instead of giving
input admittance in mhos, the input capacitance is given in farads. The
input admittance is then found when the input capacitance is multiplied
by $6.28f$ (where f is the frequency of interest). This is based on the assumption that the input admittance is primarily capacitive, and thus dependent on frequency. This is not always true for the frequency of interest.
Thus, it may be necessary to make actual tests of the MOSFET, using
complex admittance measuring equipment.

The input and output tuning circuits of an RF amplifier must perform two functions. Obviously, the circuits (capacitors and coils) must tune the amplifier to the desired frequency. The circuits must also match the input and output impedances of the MOSFET to the impedances of the source and load. Otherwise, there will be considerable loss of signal. (Note that although impedances are involved, admittances are generally used in calculations, since admittances greatly simplify the measurement and calculations.)

Finally, as in the case with a vacuum-tube amplifier or two-junction transistor common-emitter amplifier, there is some feedback between output and input of a common-source MOSFET RF amplifier. If the admittance factors are just right, the feedback will be of sufficient amplitude and of proper phase to cause oscillation of the amplifier. The amplifier is considered as *unstable* when this occurs. The condition is undesirable and can be corrected by one of two procedures: Feedback (neutralization), or by changes in the input/output tuning networks. Although the neutralization and tuning circuits are relatively simple, the equations for determining stability (or instability) and impedance matching are long and complex. Generally, *such equations are best solved by computer-aided design methods.*

In an effort to cut through this maze of information and complex equations, we shall discuss all of the steps involved in MOSFET RF design. Armed with this information, the reader should be able to interpret datasheet or test information, and use the information to design tuning networks that will provide stable RF amplification at the frequencies of interest. With each step we shall discuss the various alternative procedures and types of information available. Specific design examples are given. These examples summarize the information contained in the various steps.

Assuming that all readers may not be familiar with two-port networks, we shall start with a summary of the *y*-parameter system.

3-2.1 *y*-Parameters

Impedance (Z) is a combination of resistance (the real part) and reactance (the imaginary part). Admittance (y) is the reciprocal of impedance, and is composed of conductance (the real part) and susceptance (the imaginary part). A *y*-parameter is an expression for admittance in the form:

$$y_{is} = g_{is} + jb_{is}$$

where g_{is} is the real (conductive) part of common-source input admittance, b_{is} is the imaginary (susceptive) part of common-source input admittance.

The term $y_{is} = g_{is} + jb_{is}$ expresses the *y*-parameter in *rectangular form*.

Some manufacturers describe the y-parameters in *polar form*. For example, they give the magnitude of forward transadmittance as $|y^{fs}|$ and the angle of forward transadmittance as $\underline{/\,y_{fs}}$. Quite often, the manufacturers mix the two systems of vector algebra on their datasheets.

conversion of vector algebra forms—It is assumed that the readers are already familiar with the basics of vector algebra. However, the following notes summarize the steps necessary to manipulate vector algebra terms. This should be sufficient to perform all calculations involved in design of MOSFET RF amplifier networks.

Converting from Rectangular to Polar Form

1. Find the magnitude from the square root of the sum of the squares of the components.

$$\text{polar magnitude} = \sqrt{g^2 + jb^2}$$

2. Find the angle from the ratio of the component values.

$$\text{polar angle} = \text{arc tan}\,\frac{jb}{g}$$

   The angle is leading if the jb term is positive, and lagging if the jb term is negative.

For example, assume that the y_{fs} is given as $g_{fs} = 75$ and $jb_{fs} = 140$. This is converted to polar form by:

$$|y_{fs}| \quad \text{polar magnitude} = \sqrt{(75)^2 + (140)^2} = 159$$

$$\underline{/\,y_{fs}}\ \text{polar angle} = \text{arc tan}\,\frac{140}{75} = 61.8°$$

Converting from Polar to Rectangular Form

1. Find the real (conductive) (g) part when polar magnitude is multiplied by cosine of polar angle.
2. Find the imaginary (susceptance) (jb) part when polar magnitude is multiplied by sine of the polar angle.

If the angle is positive, the jb component is also positive. When the angle is negative, the jb component is also negative.

For example, assume that the y_{fs} is given as $|y_{fs}| = 125$ and $y_{fs} = -43°$. This is converted to rectangular form by:

$$125\cos\Theta - j125\sin\Theta = g_{fs}\ 91.7 - jb_{fs}\ 85.2$$

Adding and Subtracting Vector Algebra Forms.   Vector quantities in polar form cannot be added or subtracted directly. The values must be converted to rectangular form and then added.

Vector quantities expressed in rectangular form may be added or subtracted by adding all *g* terms, and then all *jb* terms, *algebraically.* For example, to add $g50 + jb0$, $g0 - jb120$, and $g27 + jb36$:

$$g50 + jb0$$
$$g0 \ - jb120$$
$$\underline{g27 + jb36}$$
$$g77 - jb84$$

Multiplying Vector Algebra Forms.   Vector quantities in rectangular form are multiplied as are other algebraic equations, with one exception. Since $j = \sqrt{-1}$, then $j^2 = -1$, and the j^2 term appearing in the product can be simplified in this manner. For example, to multiply $g9 + jb5$ by $g12 + jb4$,

$$g9 \ + jb5$$
$$\underline{g12 + jb4}$$
$$108 + j60$$
$$\underline{\quad + j36 + j^2 20}$$
$$108 + j96 + j^2 20$$

when the j^2 term is replaced by -1, then $108 + j96 - 20 = 88 + j96$, or $g88 + jb96$.

Vector quantities in polar form are multiplied by multiplying the magnitudes and adding the angles algebraically. This procedure is simpler than multiplying in rectangular form. Since multiplication is required in calculating MOSFET RF amplifier networks, it is often convenient first to convert any rectangular terms (that require multiplication) into polar form. As an example of polar multiplication, assume that $X \mid y_{fs} \mid 50 \ \ y_{fs} \ 20°$ is multiplied by $\mid y_{rs} \mid 32 \underline{/y_{rs}} \ -40°$.

$$50 \times 32 = 1600$$
$$20° + (-40°) = -20°$$
$$1600 \underline{/-20°}$$

Dividing Vector Algebra Forms.   Although the methods of MOSFET RF amplifier network calculations described in this book do not require division of algebraic terms, the following is included for reference. The procedure for division of polar forms is much simpler than for rectangular forms. For polar form, simply divide the magnitudes, and subtract the angles algebraically. For example, to divide $147 \underline{/\ 64°}$ by $840 \underline{/\ 35°}$:

$$\frac{147}{840} = 0.175$$
$$64° - 35° = 29°$$
$$0.175 \ \ 29°$$

the four basic y-*parameters—*Since a MOSFET can be treated as a linear active two-port network in small signal applications, all of the standard y-parameter stability criteria normally associated with two-junction transistors (and vacuum tubes) are directly applicable in RF amplifier network design. The y-parameters are extremely useful in comparing different device types, in choosing a particular configuration (common-source, common-gate, neutralized, unneutralized, cascade, etc.), and in the final design of the RF amplifier.

The y-equivalent circuit is shown in Fig. 3–10. The following is a summary of the four y-parameters of primary interest.

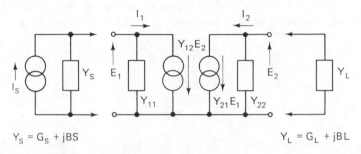

$$Y_S = G_S + jBS$$ $$Y_L = G_L + jBL$$

Fig. 3–10 MOSFET y-equivalent circuit with source and load.

Note that RF designers have traditionally used the nomenclature y_{11}, y_{12}, y_{21} and y_{22} for all active devices—two-junction transistors, integrated circuits, and other devices. Some manufacturers still follow this practice. Other manufacturers use descriptive letter subscripts y_{is}, y_{rs}, y_{fs} and y_{os} for the same parameters. Both systems are given in the following summary. (Note that the letter s refers to common-source operation.)

Input admittance, with $Y_L = $ infinity (short circuit), is expressed as:

$$y_{is} = y_{11} = g_{11} + jb_{11} = \frac{\Delta i_1}{\Delta e_1}\bigg|_{e_2 = 0}$$

Some datasheets do not show y_{is} or y_{11} at any frequency. Instead, input capacitance c_{iss} is given. If we assume that the input admittance is entirely (or mostly) capacitive (jb_{11} or jb_{is}), then the input impedance can be found when c_{iss} is multiplied by $6.28F$ ($F = $ frequency in Hz), and the reciprocal is taken. Since admittance is the reciprocal of impedance, admittance is found when c_{iss} is multiplied by 6.28F (where admittance is capacitive). For example, if the frequency is 100 MHz and the c_{iss} is 6 pF, the input admittance is $6.28 \times (100 \times 10^6) \times (6 \times 10^{-12}) \approx 3.8$ mmhos. The assumption is accurate only if the real part of y_{is} (or g_{is}) is negligible.

Figure 3–11 shows input admittance curves for a typical MOSFET.

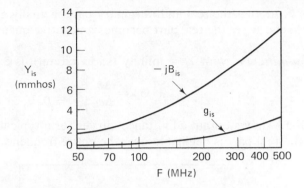

Fig. 3–11 Input admittance, $Y_{is} = g_{is} + jB_{is}$ of MFE 3007. (Courtesy Motorola)

Note that the imaginary part (jb_{is}) is the more significant factor across the entire frequency range.

Forward transmittance, with Y_L = infinity (short circuit), is expressed as:

$$y_{fs} = y_{21} = g_{21} + jb_{21} = \left.\frac{\Delta i_2}{\Delta e_1}\right|_{e_2 = 0}$$

Note that some datasheets show y_{fs} or y_{21} at some low frequency (typically 1 kHz), and then show $Re(y_{fs})$, or the real part of y_{fs}, at a higher frequency (typically 100 to 200 MHz). Other datasheets specify that g_{fs} is the real part of y_{fs}, and that the values given are for a low frequency. Then some value is given for y_{fs} at a high frequency. No matter what system is used, it is essential that the values of y_{fs} be considered at the frequency of interest.

These considerations are illustrated in Fig. 3–12 which shows more accurate and complete forward transadmittance curves for a typical

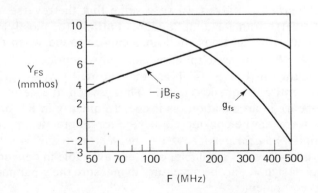

Fig. 3–12 Forward transfer admittance, $Y_{fs} = g_{fs} + jb_{fs}$ of MFE3007. (Courtesy Motorola)

MOSFET. Note that the real and imaginary parts actually cross over at about 180 MHz. Also, the real part becomes a negative quantity at about 400 MHz.

Output admittance, with Y_S = infinity (short circuit), is expressed as:

$$y_{os} = y_{22} = g_{22} + jb_{22} = \frac{\Delta i_2}{\Delta e_2}\bigg|_{e_1 = 0}$$

Figure 3–13 shows output admittance curves for a typical MOSFET. Note that the real part is negligible over the entire frequency range.

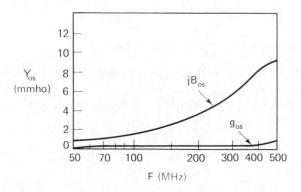

Fig. 3–13 Output admittance, $Y_{os} = g_{os} + jb_{os}$ of MFE3007. (Courtesy Motorola)

Reverse transadmittance, with Y_S = infinity (short circuit), is expressed as:

$$y_{rs} = y_{12} = g_{12} + jb_{12} = \frac{\Delta i_1}{\Delta e_2}\bigg|_{e_1 = 0}$$

Many datasheets do not list y_{rs} at any frequency. Instead, reverse-transfer capacitance C_{rss} is given. If we assume that the reverse transadmittance (or reverse-transfer admittance) is entirely (or mostly) capacitive (jb_{12}), then the reverse transadmittance can be found when C_{rss} is multiplied by 6.28F. This assumption is generally accurate in the case of y_{rs} or y_{12}, as shown in Fig. 3–14. Note that the real part of $y_{rs}(g_{rs})$ is zero across the entire frequency range. Thus, when the term $Re(y_{12})$ or $Re(y_{rs})$ appears in an equation (as it does frequently in RF design equations), the term can be considered as zero for all practical purposes.

y-parameter measurement—As can be seen thus far, y-parameter information is not always available, or is not available in convenient form. In practical design, it may be necessary to measure the y-parameters using laboratory equipment.

Note that all y-parameters are based on *ratios* of input/output current

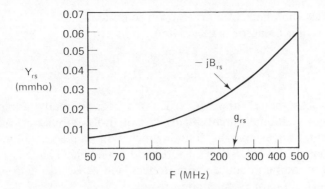

Fig. 3–14 Reverse transfer admittance, $Y_{rs} = g_{rs} + jb_{rs}$ of MFE3007. (Courtesy Motorola)

to input/output voltage. For example, y_{rs} is the ratio of output current to input voltage.

Y_{fs} and y_{os} can be measured by using signal generators, voltmeters, and simple circuits. Likewise, y_{is} and y_{rs} can be found by measuring c_{iss} and c_{rss} (using a simple capacitance meter), and then calculating the y_{is} and y_{rs} based on the frequency of interest. These procedures are described in Chap. 6.

However, more accurate results are obtained if precision laboratory equipment is used. All four y-parameters can be measured on a "General Radio Transfer Function and Immittance Bridge." A possible exception is y_{rs}, which typically is very small in relation to the other parameters. In the case of y_{rs}, it is often more practical to measure c_{rss} and multiply by $6.28F$.

The main concern in measuring y-parameters, from a practical design standpoint, is that the measurements are made under *conditions simulating those of the final circuit*. For example, the supply or drain-source voltage, gate-source voltage, bias (if any), gate 2 (if any) voltage, and operating frequency should be identical (or close) to the final circuit. Otherwise, the tests can be misleading.

3–2.2 Stability Factors

There are two factors used to determine the potential stability (or instability of MOSFETs in RF amplifiers. (Note that these same factors are used with other devices such as two-junction transistors, JFETs, and IC amplifiers.) One factor is known as the Linvill C factor; the other is the Stern k factor. Both factors are calculated from equations requiring y-parameter information (to be taken from datasheets or by actual measurement at the frequency of interest.)

The main difference between the two factors is that the Linvill C factor assumes the MOSFET is not connected to a load, while the Stern k factor includes the effect of a given or specific load.

The Linville C factor is calculated from:

$$C = \frac{y_{12}y_{21}}{2g_{11}g_{22} - R_e(y_{12}y_{21})}$$

If C is less than 1, the MOSFET is unconditionally stable. That is, using a conventional (unmodified) circuit, no combination of load and source admittances can be found which will cause oscillation. If C is greater than 1, the MOSFET is potentially unstable. That is, certain combinations of load and source admittances will cause oscillation.

The Stern k factor is calculated from:

$$k = \frac{2(g_{11} + G_S)(g_{22} + G_L)}{y_{12}y_{21} + Re(y_{12}y_{21})}$$

where G_S and G_L are source and load conductances, respectively ($G_S = 1$/source resistance; $G_L = 1$/load resistance).

If k is greater than 1, the amplifier circuit is stable (opposite from Linvill). If k is less than 1, the amplifier is unstable. In practical design, it is recommended that a k factor of 3 or 4 be used, rather than of 1, to provide a margin of safety. This will accommodate parameter and component variations (particularly with regard to band-pass response).

Note that both equations are fairly complex, and require considerable time for their solution (unless computer-aided design techniques are used). Some manufacturers provide alternate solutions to the stability and load-matching problems usually in the form of a datasheet graph. Such a graph is shown in Fig. 3–15, which is a Linvill C factor chart for a typical MOSFET. Note that the MOSFET is unconditionally stable at frequencies above 250 MHz, but potentially unstable at frequencies below 250 MHz. At frequencies below about 50 MHz, the MOSFET becomes highly unstable.

3–2.3 Solutions to Stability Problems

There are two basic solutions to the problem of unstable RF amplifiers. First, the amplifier can be *neutralized*. That is, part of the output can be fed back (after it has been shifted in phase) to the input so as to cancel oscillation. Neutralization permits the amplifier to be matched perfectly to the source and load. Such a match is known as a *conjugate match*. A perfect conjugate match means that the MOSFET input and the source, as well as the MOSFET output and the load, are *matched resistively*, and that *all reactance* is tuned out. Neutralization requires extra components, and creates a problem when frequency is changed.

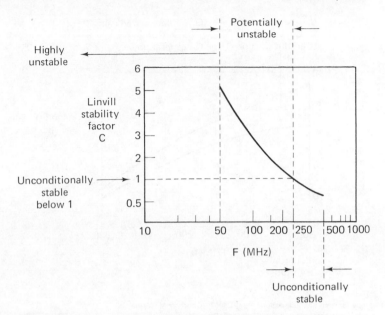

Fig. 3–15 Linvill stability factor, C, for the MFE3007 between 50 and 5000 MHz. (Courtesy Motorola)

The other solution is to *introduce some mismatch* into either the source or load tuning networks. This method, sometimes known as the Stern solution, requires no extra components, but it does produce a reduction in gain.

A comparison of the two methods is shown in Fig. 3–16. The higher gain curve represents the unilateralized (or neutralized) operation. The lower gain curve represents the circuit power gain, when the Stern *k* factor is 3.

Assume that the frequency of interest is 100 MHz. If the amplifier is matched directly to the load (perfect conjugate match), without regard to stability or with neutralization to produce stability, the top curve applies, and the power gain is about 38 dB. If the amplifier is matched to a load and source where the Stern *k* factor is 3 (resulting in a mismatch with the actual load and source), the lower curve applies, and the power gain is about 29 dB.

The upper curve of Fig. 3–16 is found by the *general power gain equation:*

$$G_P = \frac{\text{power delivered to load}}{\text{power delivered to input}} = \frac{(y_{21})^2 G_L}{(Y_L + y_{22})^2 Re(y_{11} - \frac{y_{12}y_{21}}{y_{22} + Y_L})}$$

The general power gain equation applies to circuits with no external feedback, and to circuits which have external feedback (neutralization),

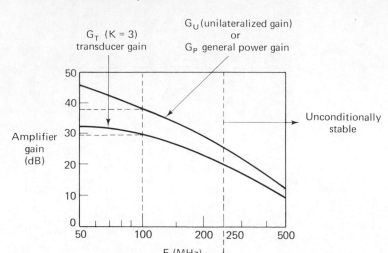

Fig. 3–16 Amplifier gain characteristics for MFE3007. (Courtesy Motorola)

provided the composite y-parameters of both MOSFET and feedback network are substituted for the MOSFET y-parameters in the equation.

The lower curve of Fig. 3–16 is found by the *transducer gain expression:*

$$G_T = \frac{\text{power delivered to load}}{\text{maximum power available from source}}$$

$$= \frac{4G_S G_L (y_{21})^2}{(y_{11} + Y_S)(y_{22} + Y_L) - (y_{12}y_{21})^2}$$

The transducer gain expression includes input mismatch. The lower curve of Fig. 3–16 assumes that the mismatch is such that a Stern k factor of 3 results. That is, the circuit tuning networks are adjusted for admittances that produce a Stern k factor of 3. The transducer gain expression considers the input and output networks as part of the source and load.

With either gain expression, the input and output admittances of the MOSFET are modified by the load and source admittances.

The input admittance of the MOSFET is given by:

$$Y_{IN} = y_{is} \ (\text{or } y_{11}) - \frac{y_{12}y_{21}}{y_{22} + Y_L}$$

The output admittance of the MOSFET is given by:

$$Y_{OUT} = y_{os} \ (\text{or } y_{22}) - \frac{y_{12}y_{21}}{y_{11} + Y_S}$$

At low frequencies, the second term in the input and output admittance equations is not particularly significant. At *VHF* (very high frequencies), the second term makes a very significant contribution to the input and output admittances.

The imaginary parts of Y_S and Y_L (B_S and B_L, respectively) must be known before values can be calculated for power gain, transducer gain, input admittance, and output admittance. Except for some very special cases, exact solutions for B_S and B_L consist of time-consuming complex algebraic manipulations.

As fairly good simplifying approximations for the equations, let $B_S \approx -b_{11}$ and $B_L \approx -b_{22}$ so that:

general power gain expression

$$G_P \approx \frac{|y_{21}|^2 G_L}{(G_L + g_{22})^2 Re\left(y_{11} - \frac{y_{12}y_{21}}{g_{22} + G_L}\right)}$$

transducer gain expression

$$G_T \approx \frac{4G_S G_L |y_{21}|^2}{[(g_{11} + G_S)(g_{22} + G_L) - y_{12}y_{21}]^2}$$

input admittance

$$Y_{IN} \approx y_{11} - \frac{y_{12}y_{21}}{g_{22} + G_L}$$

output admittance

$$Y_{OUT} \approx y_{22} - \frac{y_{12}y_{21}}{g_{11} + G_S}$$

Two other gain expressions often used in RF amplifier design are: maximum available gain (MAG), and maximum usable gain (MUG).

MAG is usually applied as the gain in a conjugately matched, neutralized circuit, and is expressed as:

$$\text{MAG} = \frac{|y_{fs}|^2 R_{in} R_{out}}{4}$$

where R_{in} and R_{out} are the input and output resistance, respectively, of the MOSFET.

An alternate MAG expression is:

$$\text{MAG} = \frac{|y_{21}|^2}{4 \, Re \, (y_{11}) Re(y_{22})}$$

where $Re(y_{11})$ is the real part (g_{is} or g_{11}) of the input admittance, and $Re(y_{22})$ is the real part (g_{os} or g_{22}) of the output admittance.

MUG is usually applied as the *stable gain* that may be realized in a *practical* (neutralized or unneutralized) RF amplifier. In a typical un-neutralized circuit, MUG is expressed as

$$\text{MUG} \approx \frac{0.4 \, y_{fs}}{6.28F \, c_{rss}}$$

where F is frequency in Hz.

3–2.4 Neutralized Solution

There are several methods for neutralization of MOSFET RF amplifiers. The most common method is the *capacitance-bridge* technique shown in Fig. 3–17. Capacitance-bridge neutralization becomes

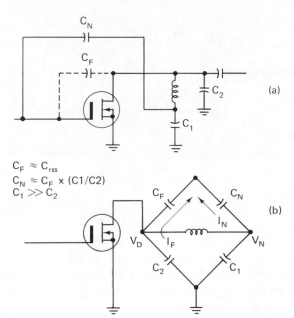

$C_F \approx C_{rss}$
$C_N \approx C_F \times (C1/C2)$
$C_1 \gg C_2$

Fig. 3–17 Capacitance-bridge neutralization circuit.

more apparent when the circuit is redrawn as shown in Fig. 3–17(b). The condition for neutralization is that $I_F = I_N$.

The equations normally used to find the value of the feedback neutralization capacitor C_N are long and complex. However, for practical work, if the value of C_1 is made quite large in relation to C_2 (at least 4 times), then the value of C_N can be found by:

$$C_N \approx C_F \frac{C_1}{C_2}$$

where $C_F = c_{rss}$ of the MOSFET.

In simple terms, the value of C_N is approximately equal to the value of c_{rss} times the ratio of C_1/C_2. For example, if c_{rss} is 0.2 pF, C_1 is 30 pF and C_2 is 3 pF, the C_1/C_2 ratio is 10, and $C_N \approx 10 \times 0.2$ pF = 2 pF.

3–2.5 The Stern Solution

A stable design with a potentially unstable MOSFET is possible without external feedback (neutralization) by proper choice of source and load admittances. This can be seen by inspection of the Stern k factor equation; G_S and G_L can be made large enough to yield a stable circuit, regardless of the degree of potential instability. Using this approach, a circuit stability factor k (typically $k = 3$) is selected, and the Stern k factor equation is used to arrive at values of G_S and G_L which will provide the desired k. Of course, the actual G of the source and load cannot be changed. Instead, the input and output tuning circuits are designed *as if the actual G values were changed.* This results in a mismatch, and a reduction in power gain, but does produce the desired degree of stability.

To get a particular circuit stability factor, the designer may choose any of the following combinations of matching or mismatching of G_S and G_L to the MOSFET input and output conductances, respectively:

G_S matched and G_L mismatched

G_L matched and G_S mismatched

both G_S and G_L mismatched

Often a decision on which combination to use will be dictated by other performance requirements or by practical considerations.

Once G_S and G_L have been chosen, the remainder of the design may be completed by using the relationships that apply to the amplifier without feedback. Power gain and input/output admittances may be computed by using the appropriate equations (Sec. 3–2.3).

simplified Stern approach—Although the above procedure may be adequate in many cases, a more systematic method of source and load admittance determination is desirable for designs that demand maximum power gain per degree of circuit stability. Stern analyzed this problem and developed equations for computing the best G_S, G_L, B_S and B_L for a particular circuit stability factor (Stern k factor). Unfortunately, these equations are very complex and quite tedious if done frequently. The complete Stern solution is best solved by computer.

Programs have been written (in BASIC) to provide essential information for MOSFETs used as RF amplifiers, including the effects of various specific sources and loads. These programs permit the designer to experiment with theoretical *breadboard* circuits in a matter of seconds.

When a Stern solution must be obtained *without the aid of a computer*, it is best to use one of the many shortcuts that have been developed over the years. The following shortcut is by far the simplest and most widely accepted, while providing an accuracy close to that of the computer solutions.

1. Let $B_S \approx -b_{11}$ and $B_L \approx -b_{22}$, as in the case of the Sec. 3–2.3 equations. This method permits the designer to closely approximate the exact Stern solution for Y_S and Y_L, while avoiding the portion of the computations that is the most complex and time consuming. Further, the circuit can be designed with tuning adjustments for varying B_S and B_L, thereby creating the possibility of experimentally achieving the true B_S and B_L for maximum gain as accurately as if all the Stern equations had been solved.

2. Mismatch G_S to g_{11} and G_L to g_{22} by an *equal ratio*. That is, find a ratio that produces the desired Stern k factor, and then mismatch G_S to g_{11} (and G_L to g_{22}). For example, if the ratio is four-to-one, make G_S four times the value of g_{11} (and G_L four times the value of G_{22}).

If the mismatch ratio R is defined as

$$R = \frac{G_L}{g_{22}} = \frac{G_S}{g_{11}}$$

then R may be computed for any particular circuit stability (k) factor using the equation:

$$R = \left(\sqrt{k \left[\frac{|y_{21}y_{12}| + Re(y_{12}y_{21})}{2g_{11}g_{22}} \right]} \right) - 1$$

As an example, assume that it is desired to mismatch input and output circuits so that there is a Stern k factor of 4, using a MOSFET that has the following characteristics: $y_{12}y_{21} = 0.545$, $g_{11} = 4.95$, $g_{22} = 0.05$, $Re(y_{12}y_{21}) = 0.258$ (all values in mmhos).

$$R = \left(\sqrt{4 \, \frac{(0.545 + 0.258)}{2(4.95)(0.05)}} \right) - 1 \approx 1.56$$

Using the value of 1.56 for R, and the equation

$$1.56 = \frac{G_S}{g_{11}} = \frac{G_L}{g_{22}}$$

then

$$G_S = (1.56)(4.95)(10^{-3}) = 7.72 \text{ mmhos}$$

and

$$R_S = \frac{1}{G_S} = 130 \text{ ohms}$$

$$G_L = (1.56)(0.05)(10^{-3}) = 0.078 \text{ mmho}$$

and

$$R_L = \frac{1}{G_L} = 12,800 \text{ ohms}$$

The shortcut Stern method may be advantageous if source and load admittances and power gains for several different values of k are desired. Once the R for a particular k has been determined, the R for any other k may be quickly found from the equation:

$$R = \frac{(1 + R_1)^2}{(1 + R_2)^2} = \frac{k_1}{k_2}$$

where R_1 and R_2 are values of R corresponding to k_1 and k_2, respectively.

   the Stern solution with datasheet graphs—It is obvious that the Stern solution, even with the shortcut method, is somewhat complex. For this reason, some manufacturers have produced datasheet graphs that show best source and load admittances for a particular MOSFET over a wide range of frequencies. Examples of these graphs are shown in Figs. 3–18 and 3–19.

   Figure 3–18 shows both the real (G_S) and imaginary (B_S) values that will produce maximum gain, but with a stability (Stern k) factor of 3, at frequencies from 50 to 500 MHz. Figure 3–19 shows corresponding information for G_L and B_L. To use these illustrations, simply select the desired frequency, and note where the corresponding G and B curves cross the frequency line. For example, assuming a frequency of 100 MHz, $Y_L = 0.35 - j2.1$ mmhos, and $Y_S = 1.3 - j4.4$ mmhos.

   If the tuning circuits are designed to match these admittances, rather than the actual admittances of the source and load, the circuit will be stable. Of course, the gain will be reduced. Use the *transducer gain expression* G_T of Sec. 3–2.3 to find the resultant power gain.

3–2.6 Single-Gate MOSFET RF Circuit Considerations

   Although stability is probably the most important factor, there are many other considerations in design of single-gate MOSFET

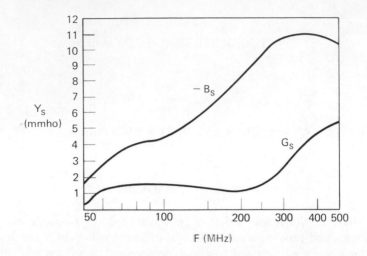

Fig. 3-18 Best source admittance, $Y_s = G_s + jB_s$. (Courtesy Motorola)

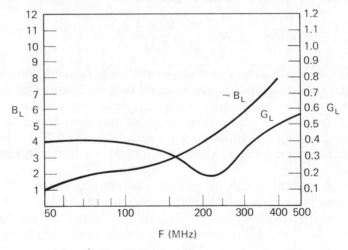

Fig. 3-19 Best load admittance, $Y_L = G_L + jB_L$. (Courtesy Motorola)

RF amplifiers. These include bias circuits, operating point, AGC, cross-modulation and intermodulation characteristics of the MOSFET and circuit. The following notes summarize the most important features of such considerations. Many of the following notes also apply to dual-gate MOSFETs. However, because of the special nature of dual-gate MOSFETs, further notes concerning their characteristics are given in later sections of this chapter.

biasing requirements—Generally, MOSFET RF amplifiers use the common-source configuration. Occasionally, common-gate will be used, particularly for RF front-end amplifiers in receivers. However, several good designs are available for common-source RF amplifiers using MOS-FETs of various manufacturers. The main objection to common-gate is the low gain at high frequencies.

Figure 3–20 shows three basic methods of dc biasing a MOSFET in the common-source circuit. Note that these methods are similar to those described in Chap. 2. All three methods are suitable for RF amplifiers.

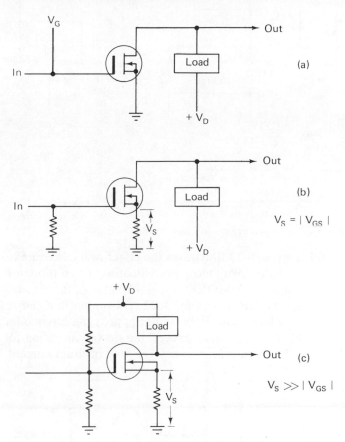

Fig. 3–20 Three basic methods for bias of a MOSFET RF amplifier in common-source configuration.

Figure 3–21 illustrates a 200 MHz common-source amplifier. This amplifier uses a modified form of the constant-current biasing arrangement shown in Fig. 3–20(c). With this modified biasing arrangement, *both the insulated gate and the case* of the MOSFET are operated at ground potential.

In any circuit, the insulated gate should always have a dc path to ground, even if the path is through a large value resistance. If the gate is allowed to "float", the resultant dc bias conditions may be unpredictable, and pos-

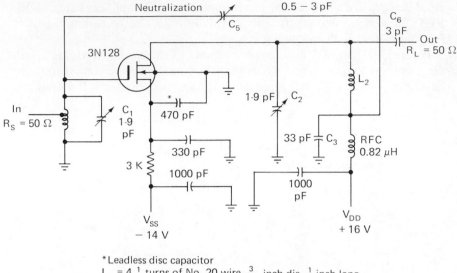

*Leadless disc capacitor

$L_1 = 4\frac{1}{2}$ turns of No. 20 wire, $\frac{3}{16}$ inch dia, $\frac{1}{2}$ inch long, tapped at 1 turn

$L_2 = 3\frac{1}{2}$ turns of No. 20 wire, $\frac{3}{8}$ inch dia, $\frac{1}{2}$ inch long

Fig. 3–21 200 MHz common-source amplifier with neutralization. (Courtesy RCA)

sibly harmful. Figure 3–22 illustrates the effect of the leakage resistances R_{L1} and R_{L2} when the insulated gate is floating. Even though the resistances are very high for a MOSFET ($\approx 10^{14}$ ohms), the resistances form a voltage divider as shown in Fig. 3–22. Assuming that the resistances are equal, the insulated gate is biased at approximately one-half of V_{DD}. This value of bias voltage may exceed the maximum rating for positive gate voltage, and may cause an excessive flow of drain current.

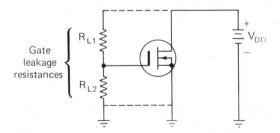

Fig. 3–22 Bias conditions for a MOSFET when the insulated gate is floating. (Courtesy RCA)

In addition to the common-source configuration, there are many MOSFET RF amplifier designs using the *cascode* configuration. As shown in Fig. 3–23, the biasing requirements for cascode are somewhat different than for a single MOSFET in common-source. In cascode, a common-

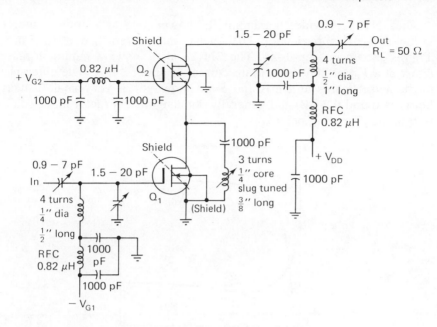

Fig. 3–23 200 MHz cascode amplifier. (Courtesy RCA)

source MOSFET is used in the lower section (or input), and a common-gate MOSFET is used in the upper section (or output). Such a circuit normally requires a negative voltage on the gate of Q_1, a positive voltage on the gate of Q_2, and approximately equal drain-to-source voltages for each MOSFET. Although the gate of Q_2 may require a positive voltage of 5 to 10V, the net gate-to-source voltage for Q_2 should be approximately 0 to −1V.

operating point considerations—As discussed in Chap. 2, the main purpose of bias is to set the operating point of the MOSFET, and to maintain that operating point. That is, a bias voltage is selected that will produce a given drain current under a given set of conditions (drain-source voltage, load, etc.). In some cases, the drain current I_D is set at the zero temperature coefficient ($0TC$) point. However, $0TC$ is not usually of great concern in RF applications. More often, an I_D is chosen to produce a given y_{fs}, power gain, noise figure, power dissipation, or some similar condition.

In many applications, a compromise between gain and available supply voltages or battery lifetime is necessary. Therefore, it is necessary to know the variation in gain and noise figure (as functions of drain voltage and drain current) before an operating point can be selected. As a point of reference, the gain and noise figure of a 3N128 MOSFET is given in the following notes.

The 3N128 provides maximum RF power gain at a drain-to-source voltage V_{DS} of approximately 20V, and a drain current I_D of about 5 mA. This is shown in Fig. 3–24. The 3N128 also shows a minimum noise figure at a V_{DS} of 20V for a drain current of about 2 mA. The difference in the noise figures at 2 mA and 5 mA, however, is very small (usually between 0 and 0.2 dB) and generally not a significant factor in the selection of the operating point.

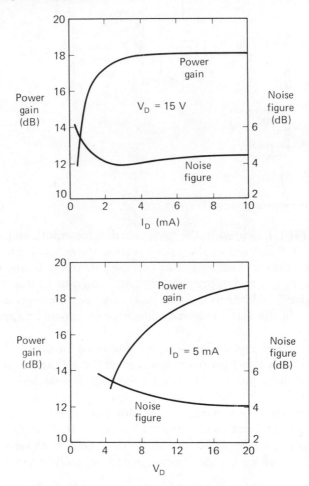

Fig. 3–24 Power gain and noise figure of 3N128 at 200 MHz as functions of I_D and V_D. (Courtesy RCA)

Although a V_{DS} of 20V represents the best value for the 3N128 in terms of both RF power gain and noise figure, this value is also the maximum V_{DS} rating for the device. Greater long-term reliability is obtained by operation of the 3N128 at a V_{DS} of 12 to 15V, rather than 20V.

For a V_{DS} of 15V and an I_D of 5 mA, the 3N128 typically provides an RF power gain of 18 dB and a noise figure of 4 dB at 200 MHz. Operation of the 3N128 at considerably lower drain currents, such as those normally used in battery-powered equipment, does not seriously affect system performance. For example, when the device is operated at a V_{DS} of 15V and an I_D of only 1 mA, the power gain and noise figure at 200 MHz are typically 15.5 dB and 4.5 dB, respectively.

Because a MOSFET is a voltage-controlled device, the performance for a given power dissipation can often be improved by operation at *high voltage* and *low current* levels. At a power dissipation of 30 mW, for example, the 3N128 typically provides a power gain of 17.3 dB and a noise figure of 3.9 dB when operated at 15V and 2 mA. At the same dissipation level, however, the power gain is reduced to 14.6 dB and the noise figure is increased to 4.7 dB when the 3N128 is operated at 6V and 5 mA.

bias for a given operating point—After the drain voltage and drain current have been selected to meet some specific design conditions, the next step is to provide a bias that will produce the desired operating point. For example, assume that V_D is to be 15V, and I_D is to be 5 mA.

A gate voltage V_G of between -0.5 and $-2V$ is normally required to bias a 3N128 for operation at $V_D = 15V$ and $I_D = 5$ mA.

If a fixed-bias circuit, as shown in Fig. 3–20(a) is used, the value of the gate voltage V_G must be adjustable to compensate for variations in individual MOSFETs. For the source-resistance bias circuit shown in Fig. 3–20(b), the value of the biasing resistor should be 200 ohms (1V/5 mA = 200 ohms). The source-resistance circuit will limit the variations in current among the difference devices to approximately 50 percent. With the constant-current bias circuit of Fig. 3–20(c), variations in current from one MOSFET to another can easily be limited to less than 10 percent.

Regardless of the bias circuit or the bias point (operating point) selected, there is no danger of thermal runaway with the 3N128 because this MOSFET has a negative temperature coefficient (at the zero-gate-bias point).

In the selection of a bias circuit for a MOSFET in which automatic gain control (AGC) is used, consideration should be given to the following. The more restrictive the tolerance imposed on operating-point current (at the Q-point), the more difficult AGC of the stage becomes. This is because the self-compensating action of the constant-current bias circuit *also resists* current changes that result from the AGC action.

AGC *systems*—When it is necessary to use AGC systems in a MOSFET RF amplifier, either of two methods may be used to reduce transistor gain for single-gate MOSFETs. (Other methods are available for dual-gate MOSFETs, as discussed in later sections.) In one single-gate method, referred to as *reverse* AGC, the reduction in gain is produced by an increase in negative gate bias. In the other single-gate method, the gain is decreased by reduction of the drain-to-source voltage.

In the reverse AGC *method*, the application of higher negative voltage to the gate reduces the drain current and the transconductance of the MOSFET. The low feedthrough capacitance of a single-gate MOSFET (typically about 13 pF for the 3N128) usually permits more than 30 dB of gain reduction at frequencies up to about 200 MHz. Substantially greater gain reduction can be had at low frequencies, or in neutralized amplifier circuits.

Gain reduction produced by the *decrease of drain-to-source voltage* is usually controlled by a variable impedance in series, or in shunt, with the MOSFET. The variable impedance may be another MOSFET or a two-junction transistor. A major disadvantage of this method is that the MOSFET feedback capacitance rises by a factor of 4 to 6 times as V_D approaches zero. This increase in capacitance reduces the AGC range obtainable, and decreases the effectiveness of a fixed neutralization network. As discussed in Sec. 3–2.4, the value of the neutralization capacitor is based on the amount of c_{rss}. If c_{rss} increases significantly, there may not be sufficient feedback through the neutralization capacitor to prevent oscillation. Another disadvantage in using reduction of V_D for AGC is that the output impedance of the MOSFET decreases with the reduction of V_D.

In a single-gate cascode circuit, AGC is produced most effectively by application of a negative voltage to the gate of the common-gate section (Q_2 of Fig. 3–23). A wide AGC range can be obtained in such a circuit. For example, in the circuit of Fig. 3–23 using 3N128 MOSFETs, gain reductions greater than 45 dB at 200 MHz, or 65 dB at 60 MHz are possible.

cross-modulation considerations—One of the major advantages of a MOSFET over a two-junction transistor, or even a JFET, is the superior cross-modulation performance. MOSFETs have lower cross-modulation distortion. Such distortion is dependent on a number of factors. However, the most significant factor is the third-order effects of the transfer characteristic. MOSFETs have a transfer characteristic (drain current I_D as a function of gate voltage V_D) that closely resembles a *quadratic curve*. If an expression is written for MOSFET I_D as a function of V_G, the third-order components are negligible over a wide range in comparison with two-junction transistor third-order components.

Cross-modulation may be defined as the transfer of information from an undesired carrier to a desired carrier. This problem is particularly important in RF amplifiers used in receivers. The first and second stages of a receiver are the most essential in consideration of cross-modulation distortion because the amplitude of the undesired carrier is insignificant in later stages as a result of the selectivity of the first stages.

Cross-modulation may be explained as follows. If a device has a *non-linear* transfer (output current/signal voltage) characteristic, and has both a desired signal V_1 and an interfering signal V_2 applied at its input, the amplification A_1 of the desired signal V_1 can be expressed as a function of the undesired signal voltage. In a practical receiver, the gain of the desired signal can be affected by the amplitude of the undesired signal, if the transfer characteristic is nonlinear.

If the transfer characteristic is either linear or quadratic, the gain A_1 is independent of the input voltage amplitude. If such is the case, A_1 is also independent of the amplitude of the interfering signal. However, if the transfer characteristic deviates from either the linear or quadratic form, the amplification of the desired signal depends on the magnitude of V_2. Thus, if V_2 is amplitude-modulated, A_1 varies with the modulation on V_2, and produces cross-modulation when the amplitude modulation information is transferred from V_2 to V_1 through gain fluctuations of A_1.

As a point of reference, at maximum gain, the cross-modulation distortion of the 3N128 is approximately one-tenth that of most two-junction transistors. However, *cross-modulation distortion of* MOSFETs *is affected by the amount and type of* AGC *applied*. This is shown in Figs. 3–25 and 3–26. The procedures for measurement of cross-modulation distortion are described in Chap. 6.

For a MOSFET connected in a circuit similar to that of Fig. 3–21, the cross-modulation distortion increases when reverse AGC is applied. If AGC action is produced by reduction of the drain-to-source voltage, the distortion is also increased, but to a lesser extent.

For MOSFETs connected in cascode similar to that of Fig. 3–23, the effect of reverse AGC on cross-modulation distortion is reduced when the AGC is applied to the gate of the common-gate stage (Q_2). Application of reverse bias to the gate of the common-source stage (Q_1) results in cross-modulation performance similar to that of a single triode-connected stage.

intermodulation considerations—Another major advantage of MOS-FETs is the superior intermodulation performance. If an RF amplifier, or mixer, has third-order nonlinearities in its transfer characteristic, two strong signals at the input of the device will produce several new frequencies. Two of these new frequencies are $2f_1 - f_2$ and $2f_2 - f_1$. For example, if f_1 is 175 MHz and f_2 is 200 MHz, there will be strong signals at 150 MHz ($2 \times 175 - 200 = 150$) and 225 MHz ($2 \times 200 - 175 = 225$).

As a point of reference, the intermodulation distortion levels of a 3N128 are approximately two to five times lower than that of most two-

Desired frequency = 200 MHz   Triode: $V_D = 15$ V, $I_D = 5$ mA
Interfering frequency = 150 MHz   Cascode: $V_D = 20$ V, $I_D = 10$ mA

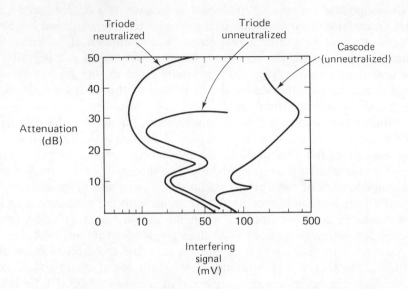

Fig. 3–25 Cross-modulation distortion as a function of the attenuation produced by reverse AGC for 3N128. Desired frequency = 200 MHz; Triode: $V_D = 15$V, $I_D = 5$mA; Interfering frequency = 150 MHz; Cascode: $V_D = 20$V, $I_D = 10$ mA. (Courtesy RCA)

Desired frequency = 200 MHz
Interfering frequency = 150 MHz $V_{DS} = 15$ V, $I_D = 5$ mA

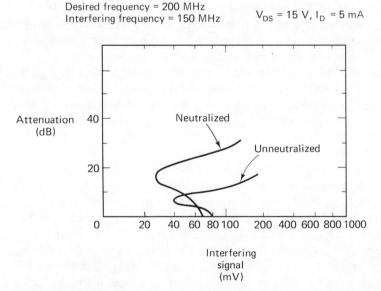

Fig. 3–26 Cross-modulation distortion as a function of the attenuation produced when AGC is accomplished by a reduction of drain-to-source voltage for 3N128. Desired frequency = 200 MHz; $V_{DS} = 15$V, $I_D = 5$ mA; Interfering frequency = 150 MHz. (Courtesy RCA)

junction transistors. Typical intermodulation distortion data for the 3N128 is shown in Fig. 3–27. The procedures for measurement of intermodulation distortion are described in Chap. 6.

The receiver used to measure distortion in the measurements taken to produce the data of Fig. 3–27 is tuned to 150 MHz. During test, the two signals are applied to the MOSFET input. The MOSFET output is fed to the receiver. Bias voltages are applied to the MOSFET to produce the various values of I_D.

Interfering voltages required to produce 2.4
microvolts at 150 MHz, with V_D = 16.5 V

I_D (mA)	f_1 (175 MHz) (mV)	f_2 (200 MHz) (mV)
10	18	18
10	7	150
5	15	15
5	3.5	150
5	30	3.5
2.5	19	21

Fig. 3–27 Intermodulation distortion data for 3N128. (Courtesy RCA)

When no signals are applied (that is, the amplitudes of signals f_1 and f_2 are both $0V$), the RF indicator of the receiver indicates an equivalent input noise level of 2.4 microvolts. The signals f_1 and f_2 are gradually increased in amplitude until the reading on the RF indicator is one microvolt above the noise level (3.4 microvolts total), indicating that 2.4 microvolts of 150-MHz signal are being produced by the interaction of f_1 and f_2 (that is, $2f_1 - f_2 = 150$ MHz). Figure 3–27 shows the I_D levels of the MOSFET and the amplitude of the f_1 and f_2 signals required to produce an output, at 150 MHz, of 2.4 microvolts (which corresponds to one microvolt above the input noise level).

3–2.7 Systematic MOSFET RF Amplifier Design Procedure

The first step in any amplifier design is to decide on an operating point. In the case of a MOSFET, this means setting the bias to produce a given I_D (for gain, noise figure, etc.) as described in Chap. 2 and Sec. 3–2.6. Once the bias and operating point have been selected, the tuning and matching networks must be designed. Basically, the steps are:

1. Determine the potential instability of the MOSFET. This involves extracting the y-parameters of the MOSFET from the datasheet, or determining the y-parameters from actual test. Next, plug the y-parameters into the Linvill C and/or Stern k equations to find potential

stability or instability. Use the Linvill C factor where source and load impedances are not involved (or known). Use the Stern k factor when load and source impedances are known. As a practical matter, it is usually more convenient to go directly to the Stern k factor, since this serves as a starting point if the circuit must be modified to produce stability.

2. If the MOSFET is not unconditionally stable, decide on a course of action to insure circuit stability. Usually, this involves going to neutralization or mismatching input/output tuning circuits. Mismatching is, by far, the most popular course of action. However, examples of both methods are described in later paragraphs of this section. If the MOSFET is unconditionally stable, without neutralization or mismatch, the design can proceed without fear of oscillation. Under these circumstances, the usual object is to get maximum gain by matching the tuning circuits to the actual source and load.

3. Determine source and load admittances. Source and load admittance determination is dependent on gain and stability considerations, together with practical circuit limitations. If the MOSFET is potentially unstable at the frequency of interest and with actual source and load impedances, then a source and load that will guarantee a certain degree of circuit stability must be used. This involves the Stern solution described in Sec. 3–2.5. If optimum source and load impedances are given by a manufacturer's datasheet, use these as a first choice. As a second choice, use computer-aided design techniques to get a Stern solution for the desired stability and gain. If neither of these is available, use the short-cut Stern technique of Sec. 3–2.5. Note that it is a good idea to check circuit stability (Stern k) factor, even when an unconditionally stable MOSFET has been found by the Linvill C factor. A MOSFET may be stable without a load, or with certain loads, but not stable with some specific load.

   Once the best source and load admittances have been selected, verify that the required gain will be available. In practical terms, it is possible to mismatch almost any MOSFET RF amplifier sufficiently to produce a stable circuit. However, the resultant power gain may be below that required. In that case, a different MOSFET must be used (or a lower gain accepted).

4. Design appropriate networks (input/output tuning circuits) to provide the desired (or selected) source and load admittances. First, the networks must be resonant at the desired frequency. (That is, inductive and capacitive reactance must be equal at the selected frequency.) Second, the network must match the MOSFET to the load and source. (That is, resistances must be equal, and reactance tuned out.) Sometimes, it will be difficult to achieve a desired source and load due to tuning range limitations, excess network losses, component limitations, etc. In such cases, the source and load admittances will be a compromise between desired performance and practical limitations. Generally, this involves a sacrifice of gain to get stability.

3-2.8 Design Example of 200 MHz Neutralized MOSFET RF Amplifier

Assume that the circuit of Fig. 3-21 is to be operated at 200 MHz. The source and load impedances are both 50 ohms. The characteristics of the MOSFET (taken from the datasheet) are as follows (all admittance values in mmhos):

$$y_{11} = 0.45 + j7.2,$$
$$y_{22} = 0.28 + j1.75,$$
$$y_{21} = 7.0 - j1.9,$$
$$y_{12} = 0 - j0.16.$$

Note that the real part of y_{12} is considered as zero and c_{rss} is 0.2 pF. Tuning is to be accomplished with standard 1-9 pF variable capacitors. In that it is assumed that the bias networks have been designed as described in Chap. 2, and Sec. 3-2.6, no special considerations need be given to bandwidth or selectivity in this example.

When the y-parameters are plugged into the Linvill equation, the Linvill stability factor C is over 2

$$C = \frac{|(7.0 - j1.9)(0 - j0.16)|}{2(0.45)(0.28) - Re(7.0 - j1.9)(-j0.16)} \approx 2.08$$

Thus, the MOSFET is not unconditionally stable and neutralization may be required. It is possible that a mismatch between the MOSFET input/output impedances and the 50 ohm source/load impedances would produce sufficient stability. However, to be sure, neutralization is used.

Since neutralization is used, the amplifier can be conjugately matched *for maximum gain*. With the real part of y_{12} assumed to be zero ($g_{12} = 0$), and a conjugate match, the maximum available gain expression can be used to find a MAG of about 20 dB.

$$MAG = \frac{(7.0 - j1.9)^2}{4(0.45)(0.28)} = 104 = 20.2 \text{ dB}$$

The source and load impedances must be matched to the MOSFET input and output impedances, respectively, to obtain the maximum gain. The input is matched by means of transformation (coupling autotransformer turns ratio). The output is matched by means of coupling capacitor reactance.

the source input is 50 ohms—For a conjugately matched input, the source conductance G_S and the real part of the MOSFET input admittance must *appear* to be equal. Source conductance is the reciprocal of 50 ohms, or 20 mmhos. The real part of the MOSFET input admittance

$Re(y_{11})$ is 0.45 mmho. The transformation ratio is approximately 44 (20/0.45), with a turns ratio of about 6.6 ($\sqrt{44}$). Experimentally, a turns ratio of about 4 was found to be approximately the best value. The difference results, in part, from the fact that the parallel resistance of the tank coil was not considered in the calculation.

Assuming that the input tuning capacitor is set to a value between 1 and 2 pF, the required input inductance is about 0.5 μH.

$$L(\text{in } \mu\text{H}) = \frac{2.53 \times 10^4}{\text{freq (in MHz)}^2 \times \text{capacitance (in pF)}}$$

the load output is also 50 ohms — Because the dc drain voltage must be blocked from the load, a series coupling capacitor C_S must be used. Capacitor C_S also functions to match the MOSFET output to the load. The load conductance G_L 20 mmhos, and the real part of the MOSFET output admittance $Re(y_{22})$ is 0.28 mmho. The reactance for C_S that will produce a conjugate match is found by:

$$XC_S = R_S \sqrt{\frac{R_P}{R_S} - 1}$$

where R_P is the reciprocal of the parallel MOSFET output admittance or $1/Re(y_{22})$, R_S is the reciprocal of the load conductance, $1/G_S$ (or the load impedance), and R_P is much greater than R_S.

Using these values, the reactance of C_S is:

$$XC_S = 50 \sqrt{\frac{3600}{50} - 1} \approx 420 \text{ ohms}$$

At 200 MHz, the value for C_S is found by:

$$C_S = \frac{1}{6.28 \times (200 \times 10^6) \times 420} \approx 1.9 \text{ pF}$$

Experimentally, a 3 pF capacitor was found to perform satisfactorily.

The capacitance of C_S is in parallel with the output capacitance of the MOSFET. As a guideline, the MOSFET output capacitance can be found when the imaginary part of the output admittance jb_{22} is put into the following equation:

$$\text{Parallel capacitance} = \frac{1}{6.28 \times (200 \times 10^6) \times \dfrac{1}{1.75} \text{ mmhos}} \approx 1.4 \text{ pF}$$

Note that the output admittance of the MOSFET does not truly equal y_{22}, except when the input is shorted, and y_{12} is truly zero. Thus, the jb_{22}

figure is not necessarily accurate. However, the jb_{22} figure can be used to find a trial value for design purposes in this example. If bandwidth and/or selectivity are of concern, a more exact figure for y_{22} is required. This figure can be found using the output admittance equations of Sec. 3–2.3. The problems of true input and output admittances for the MOSFET are discussed in other examples of this chapter.

In this example, a 1.4 pF output capacitance is considered to be in parallel with the 3 pF capacitance of C_S, resulting in a total capacitance of 4.4 pF. To simplify calculations, the variable tuning capacitor is considered as set to 2 pF. This produces a total parallel output capacitance of 6.4 pF.

The required output inductance is about 0.1 μH, found from the equation:

$$L(\text{in } \mu\text{H}) = \frac{2.53 \times 10^4}{(200)^2 \times 6.4} \approx 0.1 \ \mu\text{H}$$

An *approximate value for C_N is found* when c_{rss} is multiplied by the ratio of C_3 to C_2 (see Sec. 3–2.4). The value of C_2 is the total parallel output capacitance of 6.4 pF. The value of C_3 must be at least 4 times that of C_2. In this example, a standard value of 33 pF is chosen ($5 \times 6.4 = 32$). With a c_{rss} of 0.2, the value of C_N is:

$$C_N \approx 0.2 \times (33/6.4) \approx 1 \text{ pF}$$

In practice, a standard variable capacitor ($0.5 - 3$ pF) is used for C_N, making it possible to adjust the feedback (neutralization) for variations in c_{rss}. Typically, the c_{rss} for the 3N128 varies between 0.15 and 0.35 pF.

3–2.9 Design Example of 100 MHz Unneutralized MOSFET

RF Amplifier Using Simplified Stern Approach with Graphs. Assume that the circuit of Fig. 3–28 is to be operated at 100 MHz. The source and load impedances are both 50 ohms. The characteristics of the MOSFET are given in Fig. 3–11 through 3–19. It is assumed that the bias networks have been designed as described in Chap. 2 and Sec. 3–2.6. The problem here is to find the best values for C_1 through C_4, as well as L_1 and L_2.

From Fig. 3–15, the Linvill stability factor C is seen to be 2.0 at 100 MHz. Therefore, mismatching or neutralization is necessary to prevent oscillation. Mismatching is used in this example. Figure 3–16 shows that for a circuit stability (Stern k) factor of 3.0, the transducer gain will be about 29 dB. The load and source admittances for the required mismatch and gain at 100 MHz (found in Figs. 3–18 and 3–19) are:

$$Y_L = 0.35 - j2.1 \text{ mmhos}$$
$$Y_S = 1.30 - j4.4 \text{ mmhos}$$

At 100 MHz, y-parameters for the MOSFET are:

$$Y_{is} = y_{11} = 0.15 + j3.0 \qquad \text{(Fig. 3–11)}$$
$$Y_{fs} = y_{21} = 10 - j5.5 \qquad \text{(Fig. 3–12)}$$
$$y_{rs} = y_{12} = 0 - j0.012 \qquad \text{(Fig. 3–14)}$$
$$y_{os} = y_{22} = 0.04 + j1.7 \qquad \text{(Fig. 3–13)}$$

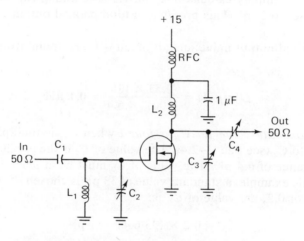

Fig. 3–28 Common-source MOSFET RF amplifier. (Courtesy Motorola)

The 50-ohm load impedance must be transformed to the optimum load for the MOSFET ($Y_L = 0.35 - j2.1$). This transformation can be performed by the network shown in Fig. 3–29(a). In effect, R_L is in series with C_4. The 50 ohms must be transformed to:

$$R_L = \frac{1}{G_L} = \frac{1}{0.35 \times 10^{-3}} \approx 2.86\text{K ohms}$$

The series capacitive reactance required for this matching can be found by:

$$XC_4 = X_{series} = R_S \sqrt{\frac{R_P}{R_S} - 1}$$

where R_P is the parallel resistance and R_S is the series resistance. Thus,

$$XC_4 = 50 \sqrt{\frac{2.86 \times 10^3}{5} - 1} \approx 372 \text{ ohms}$$

The capacitance that provides this reactance at 100 MHz is:

$$C_4 = \frac{1}{6.28FXC_4} = \frac{1}{6.28(10^8)(372)} \approx 4.3 \text{ pF}$$

The parallel equivalent of this capacitance is needed for determining the bandwidth and resonance later in the design:

$$X'C_4 = X_{parallel} = X_S \left[1 + \left(\frac{R_S}{X_S} \right)^2 \right]$$

$$= 372 \left[1 + \left(\frac{50}{372} \right)^2 \right] \approx 378$$

and the equivalent parallel capacitance is therefore

$$C'_4 \approx 4.2 \text{ pF}$$

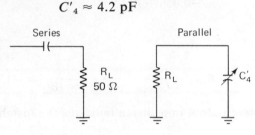

(a) Output impedance transformation

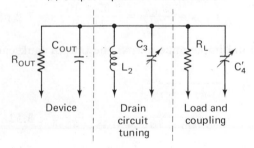

(b) Equivalent output circuit

Fig. 3-29 Impedance transformation and equivalent output circuit of common-source MOSFET RF amplifier. (Courtesy Motorola)

An equivalent circuit for the output tank, after transformation of the load, is shown in Fig. 3–29(b). Since the resistance across the output circuit is fixed by the parallel combination of R_{OUT} and R_L (after transformation), the desired bandwidth of the output tank will be determined by C_3. That is, all other factors being equal, bandwidth is determined by circuit Q. In turn, circuit Q is dependent on the ratio of reactance to resistance.

Since the resistance is fixed to match impedances, the only practical method for design of a given bandwidth is to select a proper value of C_3.

The first step is to find the actual output admittance of the MOSFET. As discussed in Sec. 3–2.8, the output admittance (Y_{OUT}) of the MOSFET will not equal y_{os} under most conditions. Only when the input is terminated in a short circuit, or the feedback admittance is zero, does Y_{OUT} equal y_{os}. When y_{rs} is not zero and the input is terminated with a practical source admittance, the true output admittance is found from the following (or from actual tests):

$$Y_{OUT} = y_{os} - \frac{y_{fs}y_{rs}}{y_{is} + Y_S}$$
$$= 0.04 + j1.7 - \frac{(10 - j5.5)(0 - j0.012)}{(0.15 + j3.0) + (1.3 - j4.4)}$$
$$= -0.066 + j2.05 \text{ mmhos}$$

therefore,

$$R_{OUT} = \frac{1}{G_{OUT}} = \frac{1}{-0.066 \times 10^{-3}} = -15.2K$$

(The negative output impedance indicates the instability of the unloaded amplifier.)

$$C_{OUT} = \frac{B_{OUT}}{6.28F} = \frac{2.05 \times 10^{-3}}{6.28(10^8)} \approx 3.2 \text{ pF}$$

Now the total impedance across the output tank can be calculated:

$$R_T = \frac{1}{G_{OUT} + G_L}$$
$$= \frac{1}{-0.066 \times 10^{-3} + 0.35 \times 10^{-3}} \approx 3.52K \text{ ohms}$$

Since the output impedance is several times higher than the input impedance of the MOSFET, amplifier bandwidth is primarily dependent on *output* loaded Q. For a bandwidth of 5 MHz (3-dB points),

$$C_T = \frac{1}{6.28 \ R_T \text{ (bandwidth)}}$$
$$= \frac{1}{6.28(3.52 \times 10^3)(5 \times 10^6)} \approx 9 \text{ pF}$$

To find C_3, subtract C_{OUT} and C'_4 from C_T, or

$$C_3 = C_T - C_{OUT} - C'_4 = 9.0 - 3.2 - 4.2 = 1.6 \text{ pF}$$

The output inductance that resonates with C_T at 100 MHz is 0.28 μH, found from the equation:

$$L_2 \text{ (in } \mu\text{H)} = \frac{2.53 \times 10^4}{(100)^2 \times 9 \text{ pF}} \approx 0.28 \, \mu\text{H}$$

Input calculations performed in a similar manner yield these results:

$$Y_S = 1.3 - j4.4 \text{ mmhos}$$
$$XC_1 = 190 \text{ ohms;} \qquad \text{therefore, } C_1 = 8.4 \text{ pF}$$
$$X'C_1 = 203 \text{ ohms;} \qquad \text{therefore, } C'_1 = 7.8 \text{ pF}$$

$$Y_{IN} = y_{is} - \frac{y_{fs} y_{rs}}{y_{os} + Y_L} = -0.25 + j4.52 \text{ mmhos;}$$

therefore,

$$R_{IN} = -4000 \text{ ohms}$$
$$C_{IN} = 7.2 \text{ pF}$$
$$R_T = 950 \text{ ohms}$$

The bandwidth of the input tuned circuit is chosen to be 10 MHz. Hence,

$$C_T = 17 \text{ pF}$$

Therefore,

$$L_1 = 0.15 \, \mu\text{H}$$
$$C_2 = 17 - 7.2 - 7.8 = 2 \text{ pF}$$

This completes design of the tuned circuits. Note that the methods described in this section provide greater accuracy than those of Sec. 3–2.8, particularly with regard to bandwidth and selectivity.

In any practical RF amplifier, it is important that the circuit be well bypassed to ground at the signal frequency, since only a small impedance to ground may cause instability or loss of gain. The bypass capacitor should be such that the reactance is about 1 to 2 ohms at the operating frequency. A 1–μF capacitor will provide less than 1–ohm reactance at 100 MHz.

3–2.10 Design Example of 100 MHz Front-End MOSFET RF Amplifier for FM Tuner

Because of their low harmonic output, MOSFETs are well-suited as front-end RF amplifiers in tuners. Spurious responses in receivers result when harmonics of an unwanted incoming signal mix with the harmonics of the local oscillator signal to produce a *difference fre-*

quency which falls with the IF *passband* of the receiver. These harmonics of unwanted signal, created in the RF amplifier, may be removed by improved filtering between the RF amplifier and the mixer. Usually, this involves double-tuning the RF stages. That is, both the RF amplifier output and mixer input (and possibly the antenna input to the RF amplifier) are tuned.

When used as RF amplifiers in receivers, the MOSFET produces an output signal that contains low levels of the harmonics of unwanted signals (for reasons discussed in Sec. 3–2.6). As a result, the need for double-tuned RF interstage transformers is reduced, and acceptable performance can generally be achieved with single-tuned circuits in both the antenna and RF interstage sections.

Figure 3–30 shows the circuit of a typical MOSFET RF amplifier used as the front-end of an FM tuner. Assume that the circuit of Fig. 3–30 is to be operated at 100 MHz, and the bias networks have been designed as described in Chap. 2 and Sec. 3–2.6. The values of circuit components obtained by means of the following design are given in the parts list of Fig. 3–30. The problem here is to find the best ratios for matching input and output circuits of the amplifier.

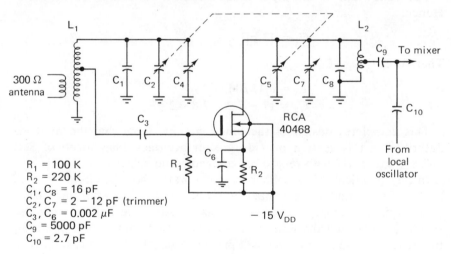

$R_1 = 100$ K
$R_2 = 220$ K
$C_1, C_8 = 16$ pF
$C_2, C_7 = 2 - 12$ pF (trimmer)
$C_3, C_6 = 0.002$ μF
$C_9 = 5000$ pF
$C_{10} = 2.7$ pF

L_1 = # 18 bare copper wire, 4 turns, $\frac{1}{4}$″ I.D., $\frac{1}{16}$″ winding length, Q_0 at 100 MHz = 120.
Tunes with 34 pF at 100 MHz.
Antenna link approximately 1 turn from ground end.
Gate tap approximately $1\frac{1}{2}$ turns from ground end.
L_2 = Same as L_1, except mixer tap at approximately $\frac{3}{4}$ turn.

Fig. 3–30 Typical FM receiver RF amplifier using 40468 MOSFET. (Courtesy RCA)

The following parameters are important in the design of the RF amplifier stage:

MOSFET (40468) parameters (at $V_{DD} = 15V$, $I_D = 5$ mA):

input resistance R_{in} .... 4500 ohms
output resistance R_{out} .... 4200 ohms
forward transadmittance y_{fs}............................... 7500 μmhos
feedback capacitance $c_{rss}(max)$ .... 0.2 pF

Mixer-stage parameters:

input resistance R_{in} .... 550 ohms
input stability $IS_{(mix)}$ 4

Coil data:

mounted unloaded Q 120
tuning capacitance C_T at 100 MHz .... 34 pF
antenna impedance... 300 ohms

Figure 3–31 shows the ac equivalent circuit for the RF stage. At resonance, this circuit reduces to the form shown in Fig. 3–32, where all impedances are referred to the gate and drain terminals of the MOSFET.

Using the maximum available gain (MAG) equation of Sec. 3–2.3, and the values for the MOSFET, MAG is

$$MAG = \frac{(7500 \times 10^{-6})^2 \times 4500 \times 4200}{4} \approx 266 \approx 24.2 \text{ dB}$$

Using the maximum usable gain (MUG) equation of the same section and the values for the MOSFET, the MUG is

$$MUG = \frac{0.4 \times (7500 \times 10^{-6})}{6.28(10^8) \times (0.2 \times 10^{-12})} \approx 23.5 \approx 13.7 \text{ dB}$$

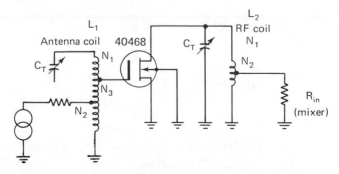

Fig. 3–31 Equivalent circuit for RF amplifier stage. (Courtesy RCA)

The total mismatch loss is called the *stability factor S*, and is equal to the *difference* (in dB) between MAG and MUG, as follows:

$$S = \text{MAG} - \text{MUG} = 24.2 - 13.7 = 10.5 \text{ dB, or } 11.3 \text{ times}$$

In the design scheme of this example, the S value is divided between the input and output circuits by means of an *input stability factor IS* and an *output stability factor OS*, as follows:

$$IS = \frac{R_{in}}{2R_1}$$

$$OS = \frac{R_{out}}{2R_2}$$

where R_1 and R_2 are the total parallel impedances of the input terminal (gate) and output terminal (drain), respectively.

These stability terms are related to the stability factor S as follows:

$$S = IS \times OS \qquad IS = \frac{S}{OS} \quad OS = \frac{S}{IS}$$

The division between IS and OS is made by means of some arbitrary choices. In this example, IS is maximized so that the signal level at the gate will be minimized. This choice necessitates matching (or nearly matching) R_{out} to its load. Therefore, the entire RF coil is used as the output load.

To achieve an impedance match between R_{out} and the load, let $OS = 1$ or unity. Then

$$IS = \frac{S}{OS}$$

or

$$IS = \frac{11.3}{1} = 11.3$$

With $OS = 1$, the value of R_2 is found by:

$$OS = \frac{R_{out}}{2R_2}, \qquad 2R_2 = \frac{R_{out}}{OS}, \qquad R_2 = \frac{\left(\dfrac{R_{out}}{OS}\right)}{2},$$

or

$$R_2 = \frac{\left(\dfrac{4200}{1}\right)}{2} = 2100 \text{ ohms}$$

With $R_2 = 2100$, the ratio of N_1/N_2 is found by:

$$\frac{N_1}{N_2} = \sqrt{\frac{IS_{(mix)} \times 2R_2}{R_{in(mix)}}} = \sqrt{\frac{4 \times 4200}{550}} \approx 5.5$$

In practice, on a four-turn coil L_2, this ratio is accomplished by making the tap at approximately 3/4-turn from the ground end.

To match the 300 ohm antenna with the input circuit, the value of R_1 should be one-half the *reflected* antenna impedance.

With $IS = 11.3$, the value of R_1 is found by:

$$IS = \frac{R_{in}}{2R_1}, \qquad 2R_1 = \frac{R_{in}}{IS}, \qquad R_1 = \frac{\left(\dfrac{R_{in}}{IS}\right)}{2},$$

or

$$R_1 = \frac{\left(\dfrac{4500}{11.3}\right)}{2} \approx 200 \text{ ohms}$$

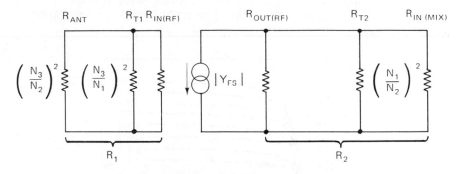

Fig. 3–32 Equivalent input R_1 and output R_2 circuit of the RF stage at resonance. (Courtesy RCA)

This value for R_1 is so much lower than R_{in} (4500), that it can be seen that the MOSFET does not load the antenna coil excessively. Also, the reflected antenna impedance is about 400 ohms, if R_1 is one-half the reflected value.

The ratio N_1/N_3 must match the reflected antenna impedance of about 400 ohms to the tuned impedance of the input coil RT_1. The ratio N_1/N_2 must match the actual antenna impedance of 300 ohms to RT_1. The value of RT_1 is dependent on the unloaded Q (given as 120), the tuning capacitance C_T (given as 34 pF), and the frequency (100 MHz). The relationship is:

$$RT_1 = \frac{Q}{6.28F \times C_T} = \frac{120}{6.28(10^8) \times (34 \times 10^{-12})} \approx 5600 \text{ ohms}$$

The N_1/N_3 transformation ratio is approximately 400/5600, or 14. This requires a turns ratio of about 3.7. The N_1/N_2 transformation ratio is approximately 300/5600, or 18.6, requiring a turns ratio of about 4.3

In practice, on a four-turn coil L_1, these ratios are accomplished by making the taps at approximately one turn from the ground end (for antenna tap N_2) and 1.5 turns from the ground end (for gate tap N_3).

3–2.11 Dual-Gate MOSFET RF Amplifier Characteristics

The dual-gate MOSFET has unique characteristics that make it well suited to certain RF amplifier applications. The advantages of dual-gate MOSFETs were discussed in Chap. 1. How these advantages can be used in RF amplifier applications is discussed in the following notes.

Basic Dual-Gate Characteristics. Figure 3–33 shows typical output characteristics for a dual-gate MOSFET (the MFE3007) connected in a common-source configuration. The curves are made with various voltages applied to gate 1, and gate 2 held fixed (at 4V). [When the MFE3007 is used as an RF amplifier, the optimum voltage for gate 2 (V_{G2S}) is 4 V.]

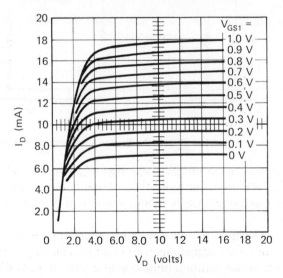

Fig. 3–33 Output characteristics of MFE3007 at $V_{G2S} = 4V$. (Courtesy Motorola)

In theory, either gate can be used for the signal input gate, since both gates show sufficient gate-to-drain transconductance (Fig. 3–34). However, due to device design and certain practical considerations, gate 1 functions best as the input gate.

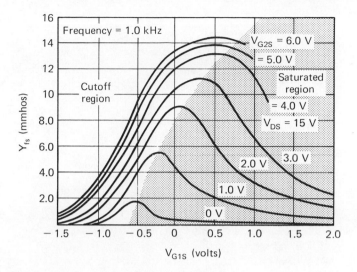

Fig. 3-34 Y_{fs1} versus bias point of MFE3007 dual-gate MOSFET. (Courtesy Motorola)

When gate 2 is used for signal injection, the channel resistance controlled by gate 1 acts similar to an unbypassed source resistance, as illustrated in Fig. 3-35. Since it is impossible to bypass this degenerative source impedance, the device gain is lowered.

Figure 3-35 also shows an equivalent representation of the device when the signal is injected at gate 1, with gate 2 bypassed to ground for alternating current. In this mode of operation, the device is equivalent to a common-source, common-gate *cascode amplifier.* The common-gate portion serves as a variable-impedance buffer stage between the input and output. The variable drain resistance in Fig. 3-35 is controlled by the dc voltage applied to gate 2. The value of the drain resistance decreases with increasing gate 2-to-source voltage. This configuration is similar to the common-cathode, common-grid cascode amplifier often used in vacuum tube circuits for high isolation between input and output Both the grounded-gate buffer and the physical separation between gate 1 and drain help achieve an extremely high degree of input-output isolation. The feedback capacitance between drain and gate 1, c_{rss}, is typically in the order of 0.02 pF.

Practical Dual-Gate RF Amplifier. Figure 3-36 shows a practical dual-gate RF amplifier operating at a frequency of 400 MHz. The circuit shown is a *simulated cascode,* with bias applied at one gate and signal applied at the other gate.

The values for C_1, C_3, C_4, L_1 and L_2 are determined as described in Sec. 3-2.9, using the same equations. In addition to designing the tuning circuits, it is necessary to select proper values for the bias networks, thus

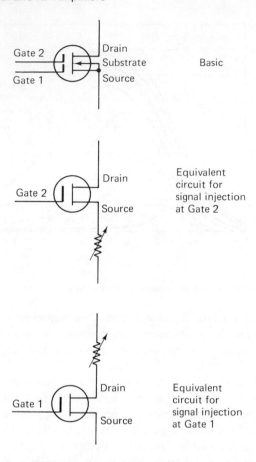

Fig. 3–35 Equivalent dual-gate MOSFET circuits with signal injection at gates 1 and 2. (Courtesy Motorola)

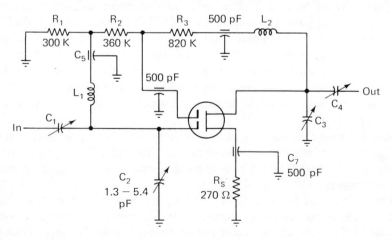

Fig. 3–36 400 MHz amplifier using dual-gate MOSFET. (Courtesy RCA)

establishing a correct operating point. The bias voltages on the two gates are set by the resistance values of the bias voltage divider network (R_1, R_2 and R_3). The amount of feedback is set by the value of R_S. The following points should be considered in selecting bias and feedback voltages.

The zero bias drain current (I_{DS}) for any MOSFET has a range of values, rather than a fixed value. The current range for the Fig. 3–36 MOSFET (3N200) is given as 0.5 mA to 12 mA, depending on production variables. Thus, a fixed bias intended to center the range of drain current at the desired level will still produce an operating drain current range of 11.5 mA. This results in a wide range of forward transconductance g_{fs}. The drain current can be regulated by applying feedback with a bypassed source resistor R_S.

A good approximation of R_S (where $I_{DQ} \geqslant I_{DS}/2$) can be calculated using the following equation, assuming that V_{G1S} versus I_{DS} is linear over the current range under consideration:

$$R_S \approx \left(\frac{1}{g_{fs}(min)}\right)\left(\frac{\Delta I_{DS}}{\Delta I_{DQ}} - 1\right)$$

where:

ΔI_{DS} is the current range given in the datasheet

ΔI_{DQ} is the desired range of operating current

$g_{fs}(min)$ is the minimum forward transconductance at 1 kHz

With the value of R_S established, the gate 1 voltage V_{G1} can be calculated from the equation:

$$V_{G1} = V_{G1S} + I_{DQ}R_S$$

where V_{G1S} is estimated by:

$$V_{G1S} \approx \frac{I_{DQ} - I_{DS}}{g_{fs}(avg)}$$

where $g_{fs}(avg)$ is the average forward transconductance.

To establish the gate 2 voltage V_{G2}, follow the same procedure described for calculating the gate 1 voltage. The recommended V_{G2S} for the 3N200 is 4V.

If gain control is desired, apply a negative-going voltage to gate 2. Because gate 2 has little control in the voltage range of +2 to +5V (for the 3N200), this characteristic may be used to produce AGC *delay* of the device in order to maintain the noise figure until the RF signal is out of the noise-level range.

Noise Figure. An important consideration for many low-level amplifiers is noise figure. It is often necessary in two-junction transistor amplifier designs to sacrifice power gain to obtain the best noise figure. Figure 3–37 shows the behavior of the noise figures and power gains of a dual-gate MOSFET (the MFE3007) versus source resistance at 100 MHz and 200 MHz.

The dual-gate MOSFET offers two important advantages over most two-junction transistors. First, very little (if any) power gain is sacrificed to get the best noise figure. Second, both noise figure and power gain are relatively independent of source resistance in the optimum region. These properties help eliminate the critical tuning and empirical adjustments usually required in front-end design. As a result, the designer has a great deal of flexibility in choosing a source impedance. In particular, as can be seen from Fig. 3–37, a 3:1 change in source resistance (300 ohms to 1000 ohms) results in only a 1-dB change in noise figure. If minimum cross modulation is a prime consideration, this 3:1 change implies a 3:1 improvement in cross modulation and total harmonic distortion for the stage.

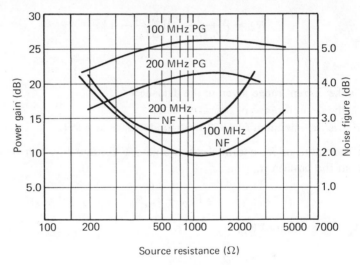

Fig. 3–37 Noise figure (NF) and power gain of the MFE3007 at 100 and 200 MHz. (Courtesy Motorola)

Automatic Gain Control for Receivers. Most RF receivers have some means of controlling the overall receiver gain to accommodate a wide range of input signal levels. This is usually done by reducing the gain of the low-level RF and IF amplifier stages.

Four different methods can be used to reduce the gain of the dual-gate MOSFET: (1) forward AGC (increasing voltage) at gate 1, (2) reverse AGC at gate 1, (3) forward AGC at gate 2, and (4) reverse AGC at gate 2. As will be shown, reverse AGC applied at gate 2 is superior to the other methods. However, each method is discussed.

Figure 3–34 shows y_{fs} as a function of the bias voltages at gates 1 and 2. Y_{fs}, and therefore the gain, can be varied by changes in the bias voltage, on either gate. Three of the AGC methods reduce gain by lowering the transconductance between gate and drain. The fourth method of gain reduction, forward AGC at gate 2, reduces the output impedance of the MOSFET.

The maximum attenuation an unneutralized RF amplifier (two-junction or MOSFET) can achieve is limited by the feedback admittance (y_{rs}). In turn, y_{rs} is dependent on c_{rss}. Since the c_{rss} of a dual-gate MOSFET is very low (even when compared to a single-gate), large gain reductions are possible with proper AGC. For example, at 200 MHz, a 50 dB gain reduction is easily obtained. When frequency is decreased, reactance increases. Thus, the same dual-gate MOSFET will provide 85 dB gain reduction at 1 MHz.

AGC applied at gate 1—With a constant gate 2 voltage, the transconductance can be reduced by varying the gate 1 voltage to either the left or the right of the peak-transconductance bias point, indicated by the shaded area in Fig. 3–34. The region to the left shows very rapid reduction in y_{fs} with decreasing bias between gate 1 and source. The sharp decrease of y_{fs} in this region is due to a rapidly approaching cutoff of that portion of the channel controlled by gate 1. This sharp cutoff introduces large third—and higher-order—nonlinearities in y_{fs} at gain reductions greater than 10 dB. These nonlinearities induce distortion and degrade cross modulation and total harmonic distortion.

Forward AGC on gate 1 is a result of reduction in y_{fs} when the bias point is shifted to the right in the shaded area of Fig. 3–34 (the saturation region). Saturation results when the majority of the channel electrons adjacent to the gate have been ionized. Therefore, to modulate the conductivity, it is necessary to ionize the electrons farther from the gate. Since the capacitive charge-inducing capability of the gate is inversely proportional to distance, greater voltage changes are necessary for equivalent changes in conductivity, and y_{fs} decreases. This method of AGC is limited to relatively small reductions because of excessive device dissipation at gain reductions greater than 10 dB. The increased device dissipation is due to the rapid increase in drain current with increasing forward bias on gate 1.

AGC applied at gate 2—The best method of AGC is reduction of the gate 2 bias voltage from its initial optimum-gain bias point (greater than 4V for the MOSFET shown in Fig. 3–34). Application of the AGC signal to gate 2 results in a *remote-cutoff characteristic*. (Remote cutoff refers to a gradual reduction in I_D with decreasing gate bias). This type of $I_D - V_G$ characteristic contrasts with the relatively abrupt cutoff of the drain current (and transconductance) when reverse AGC is applied to the gate of a single-gate MOSFET, or JFET. This sharp cutoff produces signal clipping, increasing distortion and cross modulation.

Figure 3–38 shows the gain reduction when the AGC signal is applied to gate 2. As shown, the initial gain-reduction rate is higher with a slight forward bias on gate 1 for $V_{G1S} = 0$. This is due to the faster decrease in y_{fs} with decreasing V_{G2S} in the saturation region. This faster gain reduction rate may be useful in receiver designs where the RF AGC is delayed with respect to the IF AGC, to improve overall receiver distortion.

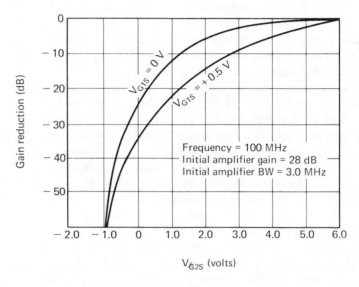

Fig. 3–38 Gain reduction versus V_{G2S} (AGC control) for dual-gate MOSFET. (Courtesy Motorola)

The fourth method of *gain reduction is by an increase in the bias voltage on gate 2*. Unlike the previous methods cited, forward AGC at gate 2 is not the result of a reduction in y_{fs}. Any increase in V_{G2S} will always result in an increase in y_{fs} regardless of the bias point. The reduction in amplifier gain when forward AGC is applied to gate 2 is a result of the lower output impedance of the device as V_{GS2} approaches the supply voltage. The value of the output impedance where this limiting occurs is about one order of magnitude less than the nominal value at V_{G2S} (4V for Fig. 3–34). Approximately 10 dB can be obtained with this type of AGC. Another disadvantage of the forward AGC method is the increase in bandwidth when the output impedance of the MOSFET is reduced.

Bandwidth and Detuning. Forward AGC is usually used in two-junction amplifier designs when cross modulation and signal-handling capability are prime considerations. With forward AGC, gain reduction also results in a considerable increase in bandwidth, and a possible shift in center frequency, unless elaborate circuit techniques are used. The increase in bandwidth is a result of the reduced output impedance of the two-junction transistor at high currents, while detuning is caused by the change in feedback and output capacitances as the output bias voltage is altered (similar to Miller effect).

When the AGC signal is applied to gate 2 of a dual-gate MOSFET, the output voltage (and thus the output capacitance c_{oss}) remains constant. The output impedance does increase slightly, but the output circuit is usually loaded sufficiently so that its effect on the bandwidth is small. For example, in a 200-MHz amplifier with 5 MHz bandwidth (−3dB points), a 50-dB gain reduction can be accompanied by only 1.25 MHz shift in center frequency and a 0.75-MHz decrease in bandwidth.

Distortion Characteristics. As discussed in Sec. 3–2.6, one of the most attractive features of the MOSFET is low distortion. The square-law transfer function of the MOSFET results in a significant improvement in signal-handling capabilities and distortion over that of two-junction transistors (which have an exponential transfer function).

A comparison of the transfer functions is shown in Fig. 3–39. These characteristics are the products of *theoretical* first-order analysis. For

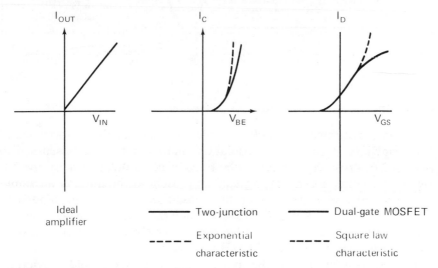

Fig. 3–39 Basic transfer characteristics of amplifiers.

the ideal linear amplifier, the second- and higher-order derivatives are zero, resulting in distortionless operation. In *practical* amplifiers (using both MOSFETs and two-junction transistors), second- and higher-order curvature does occur over the normal range of operation. However, due to the exponential nature of the two-junction transistor transfer characteristics, distortion resulting from the third- and higher-order nonlinearities is usually much more severe in two-junction amplifiers. For this reason, many present-day FM receivers are being designed with MOS-FETs in their front ends. For obvious reasons, distortion is a much greater problem in FM than AM.

Signal-Handling Capability and Cross Modulation. There are various methods of comparing the distortion characteristics of different devices. One method is the overload or signal-handling capability of the device. For the signal-handling curves shown in Fig. 3–40, overload is defined as the signal level at the input required to produce 5 percent total harmonic distortion at the output. The test circuit in this figure includes a 1-kHz common-source amplifier with an AGC signal applied at gate 2. The 1-kHz frequency is used instead of a modulated RF signal to eliminate the need for a detector in the test setup.

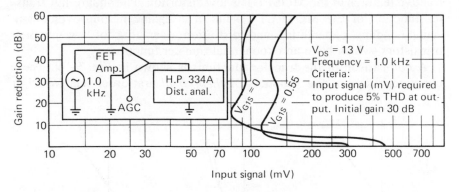

Fig. 3–40 Signal-handling capability of MFE3007. (Courtesy Motorola)

Another distortion curve often needed to evaluate the performance of RF amplifiers is the cross-modulation characteristics. As discussed in Sec. 3–2.6, cross-modulation is the transfer of modulation from one RF signal to another when both signals are present at the input of an amplifier. The precise definition for the cross-modulation index plotted in Fig. 3–41 is the level of desired signal when the undesired signal is modulated 30 percent at 1 kHz. The frequencies of the desired and undesired signals are 200 MHz, and 150 MHz, respectively.

For meaningful results, the input to the amplifier should provide no selectivity, and the desired signal should not overload the amplifier. In Fig. 3–41, measurements are made for two different V_{G1S}–bias conditions (0V and 0.55V). As the curves show, forward bias on gate 1 significantly improves cross-modulation performance.

Cross Modulation as a Function of Gain Reduction. The curves of Figs. 3–42 and 3–43 (for a 3N200 MOSFET) show the relationship of cross modulation and gain reduction. When both channels of the MOSFET are

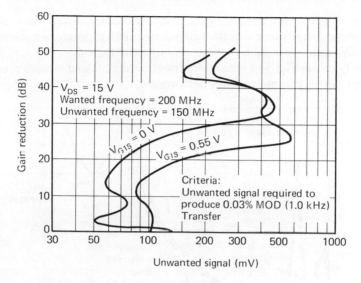

Fig. 3–41 Cross-modulation curves for MFE3007. (Courtesy Motorola)

fully conducting current, as shown by the V_{G2S}-4V curve in Fig. 3–42, the MOSFET follows an approximate square-law characteristic. If the $I_D - V_{G1S}$ curve was ideal, there would be no distortion. However, in practical cases, there is some cross modulation.

Figure 3–44 depicts three typical biasing circuits. Figure 3–43 shows the cross modulation versus gain reduction curves using the biasing circuits of Fig. 3–44. The curves are for undesired signal level required to produce a cross-modulation factor of 0.01 (considered as standard), as a

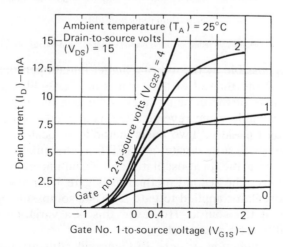

Fig. 3–42 I_D versus gate 1-to-source voltage V_{G1S} for 3N200. (Courtesy RCA)

function of gain reduction. A higher curve indicates better cross-modulation performance, since a higher curve means that greater unwanted signal is required to produce a given cross-modulation factor.

A study of the Fig. 3–43 curves shows that the biasing arrangement that provides best cross-modulation performance is circuit C. A further slight

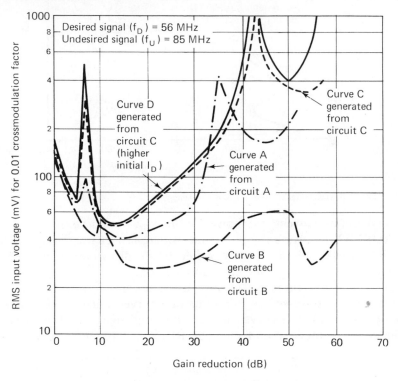

Fig. 3–43 Cross-modulation versus gain reduction using biasing circuit of Fig. 3–44. (Courtesy RCA)

improvement is possible by the use of a higher initial-operating drain current (curve D). Note that all of the circuits in Fig. 3–44 apply AGC to gate 1. Thus, none of the circuits takes advantage of the remote-cutoff characteristic available when reverse AGC is applied to gate 2.

Dual-Gate Mixer Operation. Two basic methods of local-oscillator injection can be used with a dual-gate MOSFET as the mixer. One method uses gate 1 for both the local-oscillator and signal injection, with gate 2 at some fixed bias. The second method uses gate 1 for signal injection, with the local oscillator applied to gate 2. It is also possible to inject the local oscillator at the source. However, this is a variation of the first method and will not be discussed.

local oscillator injection at gate 1—Generally, this arrangement pro-

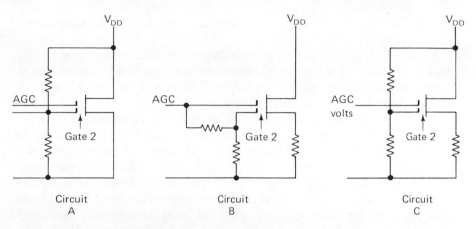

Fig. 3–44 Biasing circuits for the 3N200. (Courtesy RCA)

vides the best mixer operation. Figure 3–45 shows conversion gain versus local-oscillator injection voltage when the local oscillator is applied to gate 1 (curve *A*). Test conditions for the mixer curve are: signal frequency = 200 MHz, local-oscillator frequency = 230 MHz, intermediate frequency = 30 MHz, $V_{DS} = 15V$, $V_{G1S} = 0.5V$, and $V_{G2S} = 4V$ (ac bypassed to ground).

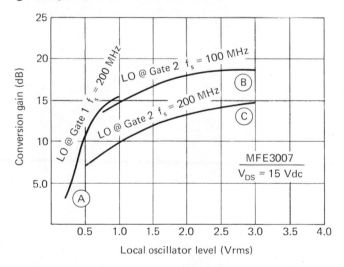

Fig. 3–45 Conversion power gain versus local oscillator injections for MFE3007. (Courtesy Motorola)

As can be seen, the conversion gain reaches a maximum of 15 dB for a local-oscillator voltage of about 1V. For injection greater than 1V, the MOSFET begins to saturate on the positive peaks of the local-oscillator voltage. In addition to limiting the conversion gain, this clipping action also increases spurious response.

local oscillator injection at gate 2—When the local oscillator is applied to gate 2, somewhat higher injection levels are required for conversion gains comparable to injection at gate 1. The reason for this is the lower gate-to-drain transconductance of gate 2.

In spite of the higher injection voltage required, injecting the local oscillator at gate 2 may serve well in mixer applications where it is desirable to *isolate the local oscillator* from both input and output networks.

Figure 3–45 also shows conversion gain versus local oscillator injection for gate 2 with signal input frequencies of 100 MHz (curve *B*) and 200 MHz (curve *C*). The local oscillator frequencies are 130 MHz and 230 MHz, respectively. In both cases, the 30-MHz difference frequency serves as the IF. The test conditions for both frequencies are $V_{DS} = 15V$, G_{G1S} and $V_{G2S} = 0.3$ to $0.5V$.

4. PRACTICAL MOSFET APPLICATIONS

This chapter is devoted to additional applications of discrete MOSFETs. These applications supplement the basic MOSFET amplifier information described in Chap. 2. Specifically, this chapter covers oscillators, choppers, attenuators, voltage variable resistors, and constant current applications.

To make full use of MOSFET advantages, and to avoid MOSFET limitations, the basic principles of the circuit are discussed briefly. These discussions are slanted as to how MOSFETs fit into the particular circuit.

4-1. OSCILLATORS

The characteristics of MOSFETs are very similar to those of vacuum tubes. Thus, practically any of the classic vacuum-tube oscillators can be designed with a MOSFET. For this reason, we will not discuss the details for all oscillator circuits. Instead, we will summarize the basic MOSFET oscillator circuits and discuss how oscillator design problems relate to MOSFET characteristics. Following this will be specific design examples of MOSFET oscillators, both RF and audio.

4-1.1 Basic MOSFET Oscillator Circuits

Figure 4-1 shows two arrangements of the *Hartley oscillator* circuit. The circuit of Fig. 4-1(a) uses a bypassed source resistor to provide proper operating conditions. The circuit of Fig. 4-1(b) uses a gate-leak resistor and biasing diode. The amount of feedback in either circuit is dependent on the position of the tap on the coil. Too little feedback results in a feedback signal voltage at the gate insufficient to sustain oscillation. Too much feedback causes the impedance between source and drain to become so low that the circuit is unstable. Output from the

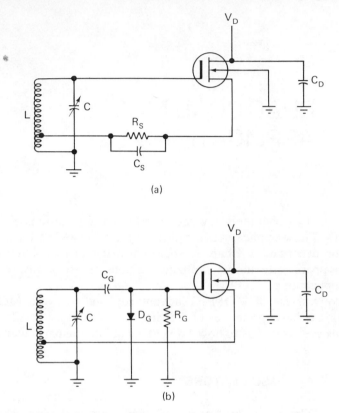

Fig. 4–1 Basic MOSFET Hartley oscillator circuits.

Hartley circuits can be obtained through inductive coupling to the coil, or through capacitive coupling to the gate.

One problem common to all oscillators, including all types of MOSFET oscillators, is the *class of operation*. If an oscillator is biased class *A* (with some I_D flowing at all times), the output waveform will be free from distortion, but the circuit will not be efficient. That is, power output will be low in relation to power input. For the purposes of calculation, input power for a MOSFET oscillator can be considered as the product of I_D and drain voltage. Class *A* oscillators are usually not good for RF, and are generally limited to those applications where a good waveform is the prime consideration.

A class *C* oscillator (where I_D is cut off by feedback greater than V_P) is far more efficient. It cuts both power and heat requirements. Since at radio frequencies, the waveform is usually not critical, class *C* is in common use for RF circuits.

In the design examples described in later paragraphs of this section, the class of operation is set by the amount of feedback, rather than by the bias point. That is, the MOSFET is biased for an optimum operating point, and then feedback is adjusted for the desired class of operation.

Figure 4–2 shows the MOSFET used in two forms of the *Colpitts oscillator circuit*. These circuits are more commonly used in VHF and UHF equipment than the Hartley circuits because of the mechanical difficulty involved in making the tapped coils required at high frequencies. Feedback is controlled in the Colpitts oscillator by the ratio of capacitance C' to C''.

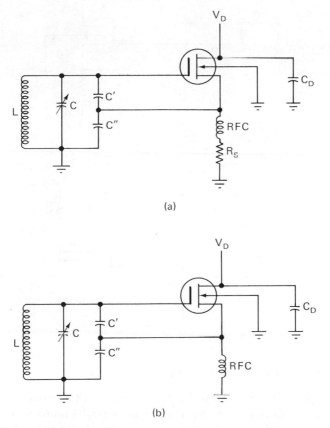

(a)

(b)

Fig. 4–2 Basic MOSFET Colpitts oscillator circuits.

The main concern in any oscillator design is that the MOSFET will oscillate at the desired frequency, and will produce the desired voltage or power. In comparison to two-junction transistors, MOSFETs have the disadvantage of lower power output. At best, with present-day MOSFETs, the power output is in the order of a few milliwatts.

The MOSFET also operates efficiently in *crystal oscillator* circuits such as the *Pierce-type oscillator* shown in Fig. 4–3. The Pierce-type oscillator is very popular because of its simplicity and minimum number of components. At frequencies below 2 MHz, a capacitive voltage divider may be required across the crystal. The connection between the voltage-divider capacitors must be grounded so that the voltage developed across the capacitors is 180° phase inverted.

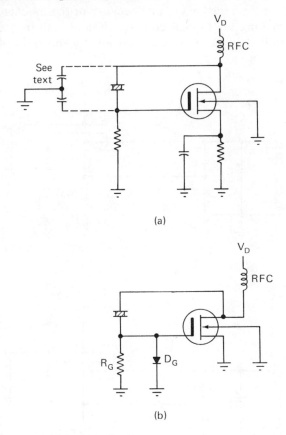

(a)

(b)

Fig. 4-3 Basic MOSFET Pierce-type oscillator circuits.

It is frequently desirable to operate crystals in communications equipment at their *harmonic or overtone frequencies.* Figure 4–4 shows two circuits designed for overtone operation. Additional feedback is obtained for the overtone crystal by means of a capacitive divider as the tank bypass. Most third-overtone crystals operate satisfactorily without this additional feedback, but this extra feedback is required for the fifth and seventh harmonics. The tank in Fig. 4–4 is not fully bypassed, and thus produces a voltage that aids oscillation. The crystal in both circuits is connected to the junction of the two capacitors C'_D and C''_D. The ratio of these capacitors should be approximately 1:3.

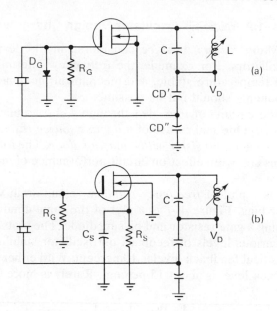

Fig. 4-4 MOSFET crystal oscillator circuits permitting operation at overtone or harmonic frequencies.

The circuit of Fig. 4–5 operates well *at low-frequencies*. The crystal is located in the feedback circuit between the sources of the two MOSFETs and operates in the series mode. Capacitor C_2 is used for precise adjustment of the oscillator frequency. A reduction in C_2 capacitance increases the frequency slightly.

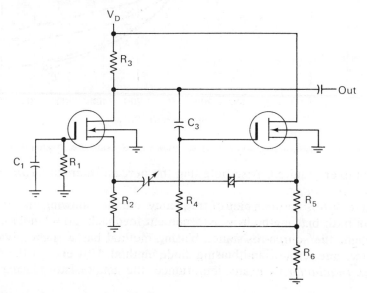

Fig. 4-5 Low-frequency MOSFET crystal oscillator.

4–1.2 MOSFET Oscillator Design Characteristics

Many factors must be considered in the design of stable MOSFET oscillators. For example, the frequency determining components must be temperature stable, and mechanical movement of the individual components should not be possible.

In each of the circuits of Figs. 4–1 through 4–4, two biasing arrangements are shown. One makes use of a *bypass source resistor,* the other arrangement uses a *gate resistor with a biasing diode.* The following notes and illustrations show the effect on circuit performance of these two biasing methods.

Figure 4–6 is a plot of *frequency stability* versus drain voltage using source-resistor bias. Figure 4–7 is a plot of the same conditions but for the circuits using a gate resistor and biasing diode. The plots show clearly the effect of various levels of feedback on oscillator stability. Note that the lowest practical feedback level is 10 percent. With either bias method, the best feedback level is about 15 percent. Rarely is more than 25 to 30

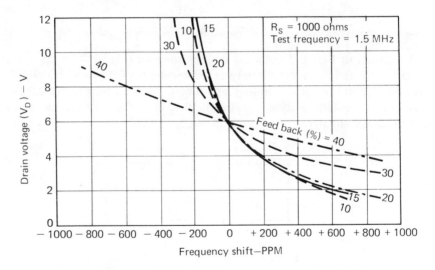

Fig. 4–6 Frequency stability versus drain voltage for circuits with source-resistor bias. (Courtesy RCA)

percent ever required. Also note that the percent refers to *feedback versus output voltage.*

Figure 4–8 is another plot of frequency stability showing the performance of both bias methods at a 15 percent feedback ratio. Under normal operation, the source-resistance biasing method has a slight advantage over the gate resistor and biasing diode method. However, if the output voltage *regulation* is of any importance, the gate-resistor-biasing-diode

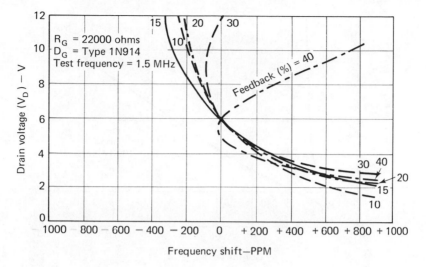

Fig. 4–7 Frequency stability versus drain voltage for circuits biased with gate resistor and biasing diode. (Courtesy RCA)

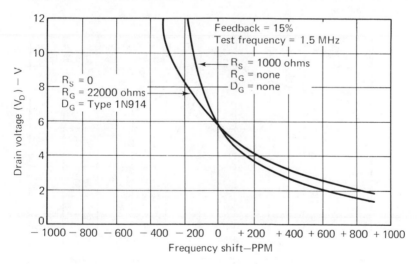

Fig. 4–8 Comparison curves of 15 percent feedback for source-resistor and gate-resistor-biasing-diode methods. (Courtesy RCA)

method is superior. Figure 4–9, a plot of output voltage versus drain voltage for both biasing methods, shows this condition.

Figures 4–10 and 4–11 show how frequency is affected when MOSFET oscillators are loaded. Note that with source-resistor bias the oscillator stops functioning at lower load levels than when bias is provided by the resistor and biasing diode.

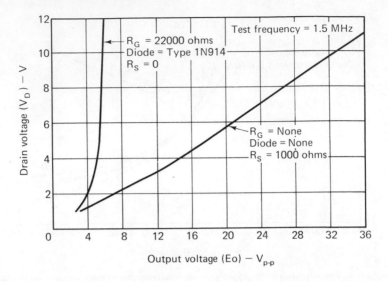

Fig. 4-9 Output voltage versus drain voltage for the source-resistor and gate-resistor-biasing-diode methods. (Courtesy RCA)

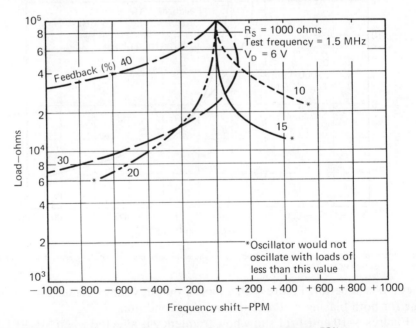

Fig. 4-10 Frequency shift versus load for the source-resistor bias method. (Courtesy RCA)

Figure 4-12 is another plot of frequency stability, showing the performance of both bias methods operating with different load at a 15 percent feedback ratio. Note that there is very little frequency shift with a 15 percent feedback ratio.

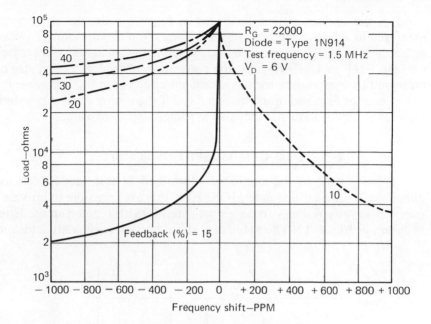

Fig. 4–11 Frequency shift versus load for the gate-resistor and biasing diode method.

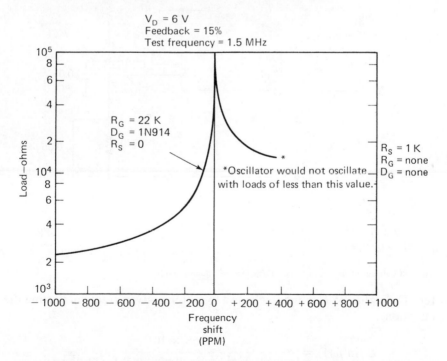

Fig. 4–12 Fifteen percent feedback comparison curves for both biasing methods with circuits loaded. (Courtesy RCA)

All of the characteristics illustrated in Figs. 4–6 through 4–12 were measured at a frequency of 1.5 MHz, using a tuning capacitor C valued at 2 pF per meter. If the operating frequency of the oscillators is increased into the VHF or UHF regions, the percentage of feedback must also be increased to compensate for the additional circuit loading. Likewise, the percentage of feedback must be increased if the tuning circuits are made high-C through substantial increase in tuning capacitor value.

4–1.3 Dual-Gate MOSFET Oscillators

Any of the oscillators described in this section can use either single-gate or dual-gate MOSFETs. It is also possible to provide a *gated* or *keyed oscillator* using a dual-gate MOSFET. Such an oscillator is shown in Fig. 4–13. The value of C should be approximately 2 pF/me-

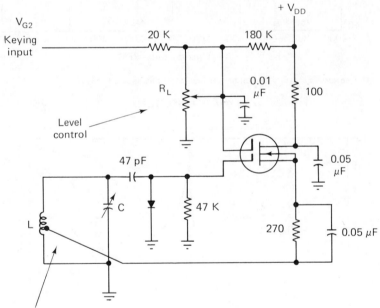

Tap at 20% of turns from cold end of coil
C = 2 pF/meter

Fig. 4–13 Gate-keyed oscillator using dual-gate MOSFET. (Courtesy RCA)

ter. The value of L is then selected to match a given frequency using the equation:

$$L \text{ (in } \mu\text{H)} = \frac{2.53 \times 10^4}{\text{freq (in MHz)}^2 \times \text{capacitance } C \text{ (in pF)}}$$

4–1.4 MOSFET Crystal Oscillator

Figure 4–14 is the working schematic of a crystal-controlled oscillator. This circuit is one of the many variations of the Colpitts oscillator. However, the output frequency is fixed, and is controlled by the crystal. The circuit can be used over a narrow range by L_1 (which is slug-tuned). For maximum efficiency, the resonant circuit (C_1, C_2, L_1 and the MOSFET output capacitance) should be at the same frequency as the crystal. If reduced efficiency is acceptable, the resonant circuit can be at a higher frequency (multiple) of the crystal frequency. However, the resonant circuit should not be operated at a frequency higher than the fourth harmonic of the crystal frequency.

bias circuit—The bias circuit components, R_1, R_2 and R_S, are selected using the procedures described in Chap. 2. That is, a given I_D is selected, and the bias components are chosen to produce that I_D under no-signal conditions. The bias circuit is calculated and tested on the basis of normal operating point, even though the circuit will never be at the operating point. A feedback signal is always present, and the MOSFET is always in a state of transition.

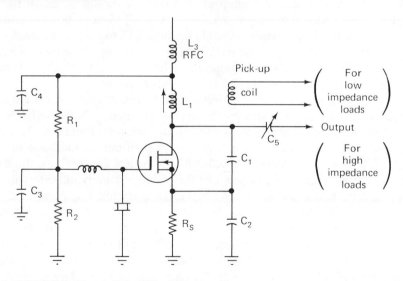

Fig. 4–14 MOSFET crystal-controlled oscillator design example.

If temperature stability is of prime importance, the I_D should be at the $0TC$ point. If power output is the main concern, the I_D should be set at a value to produce the required power output. With the correct bias-feedback relationship, the output power of the oscillator will be about 0.3 times the input power (which is drain voltage or supply voltage times I_D). Typically, the voltage drop across L_1 and L_3 is very small, so that the drain voltage equals the supply voltage. Thus, to find a correct value of I_D for a given power output and supply voltage, divide the desired output into 0.3 to find the required input power. Then divide the input power by the supply voltage to find I_D.

feedback signal—The signal output appears at the drain terminal. With the proper bias-feedback relationship, the output signal is about 80 percent of the supply voltage. The amount of feedback is determined by the ratio of C_1 and C_2. For example, if C_1 and C_2 are of the same value, the feedback signal is one-half of the output signal. If C_2 is made about three times the value of C_1, the feedback signal is about 0.25 of the output signal voltage.

It may be necessary to change the value of C_1 in relation to C_2, to obtain a good bias-feedback relationship. For example, if C_2 is decreased in value, the feedback increases and the oscillators operate nearer the class C region. An increase in C_2, with C_1 fixed, decreases the feedback and makes the oscillator operate as class A. Keep in mind that any change in C_2 (or C_1) will also affect frequency. Thus, if the C_2/C_1 values are changed, it will probably be necessary to change the value of L_1.

As a first trial value, the amount of feedback should be equal to, or greater than, maximum V_P [or maximum $V_{GS(off)}$]. Under normal conditions, such a level of feedback should be sufficient to overcome the fixed bias (set by R_1 and R_2) and the variable bias set by R_S. As shown in Figs. 4–6 through 4–12, feedback is generally within the limits of 10 to 40 percent, with best stability in the 15 to 25 percent range.

frequency—Frequency of the circuit is determined by the resonant frequency of L_1, C_1 and C_2, and by the crystal frequency. Note that C_1 and C_2 are in series, so that the total capacitance must be found by the conventional series equation. Also note that the output capacitance of the MOSFET must be added to the value of C_1. At low frequencies, the output capacitance can be ignored since the value is usually quite low in relation to a typical value for C_1. At higher frequencies, the value of C_1 is lower, so the output capacitance becomes of greater importance. For example, if the output capacitance is 5 pF at the frequency of interest, and the value of C_1 is 1000 pF or larger, the effect of the output capacitance will be small. (MOSFET output capacitance can be considered as being in parallel with C_1.) If the value of C_1 is lowered to 5 pF, the parallel output capacitance will double the value. Thus, the output capacitance must be included in the resonant-frequency calculation.

MOSFET output capacitance is not always listed on datasheets. When it is, output capacitance is listed as c_{oss}. Actually, the capacitance presented by the MOSFET from drain to source (across C_1) is composed of both c_{oss} and c_{rss}. However, since c_{rss} is always very small in relation to c_{oss} for any MOSFET, c_{rss} can be ignored in this case.

When c_{oss} is not available on datasheets, it is possible to calculate an approximate value of c_{oss} from y_{os}. The imaginary part of y_{os} (jb_{os} or j_{b22}) represents susceptance, which is the reciprocal of reactance. Thus, to find the reactance presented by the drain-source terminals of the MOSFET at the datasheet frequency, divide jb_{os} into one. Then find the capacitance that will produce such reactance, using the equation:

$$C = \frac{1}{6.28F \ X_C}$$

where C is capacitance, F is frequency, and X_C is capacitive reactance found as the reciprocal of jb_{os}.

Of course, this method assumes that the jb_{os} reactance is capacitive, and that the capacity remains constant at all frequencies (at least it is the same for the datasheet frequency and design frequency).

Capacitor C_1 can be made variable. However, it is generally easier to make L_1 variable, since the tuning range of a crystal controlled oscillator is quite small.

Typically, the value of C_2 is about three times the value of C_1 (or the combined values of C_1 and the MOSFET output capacitance, where applicable). Thus, the signal voltage (fed back to the MOSFET source terminal) is about 0.25 of the total output signal voltage (or about 0.2 of the supply voltage, when the proper bias-feedback relationship is established).

resonant circuit—Any number of L and C combinations can be used to produce the desired frequency. That is, coil can be made very large or very small, with corresponding capacitor values. Often, practical limitations are placed on the resonant circuit (such as available variable inductance values). In the absence of such limitations, and as a starting point for resonant-circuit values, the capacitance should be 2 pF per meter. For example, if the frequency is 30 MHz, the wavelength is 10 meters, and the capacitance should be 20 pF. Wavelength in meters is found by the equation:

$$\text{wavelength} = \frac{300}{\text{frequency (MHz)}}$$

At frequencies below about 1 to 5 MHz, the 2 pF/meter guideline may result in very large coils to produce the corresponding inductance. If so, the 2 pF/meter can be raised to 20 pF/meter.

output circuit—Output to the following stage can be taken from L_1 by means of a pick-up coil (for low-impedance loads) or coupling capacitor (for high impedance loads). Generally, the most convenient output scheme is to use a coupling capacitor (C_5), and make the capacitor variable. This makes it possible to couple the oscillator to a variable load (a load that changes impedance with changes in frequency).

crystal—The crystal must, of course, be resonant at the desired operating frequency (or a submultiple thereof, when the circuit is used as a multiplier). Note that efficiency (power output in relation to power input) of the oscillator is reduced when the oscillator is also used as a multiplier. The crystal must be capable of withstanding the combined dc and signal voltages at the MOSFET gate. As a rule, the crystal should be capable of withstanding the full supply voltage, even though the crystal will never be operated at this level.

bypass and coupling capacitors—The values of bypass capacitors C_3 and C_4 should be such that the reactance is 5 ohms or less at the crystal operating frequency. A higher reactance (200 ohms) could be tolerated; however, due to the low crystal output, the lower reactance is preferred.

The value of C_5 should be approximately equal to the combined parallel output capacitance of the MOSFET and C_1. Make this the mid-range value of C_5, if C_5 is variable.

radio-frequency chokes—The values of radio-frequency chokes (RFC) L_2 and L_3 should be such that the reactance is between 1000 and 3000 ohms at the operating frequency. The minimum current capacity of the chokes should be greater (by at least 10 percent) than the maximum anticipated direct current. Note that a high reactance is desired at the operating frequency. However, at high frequencies, this can result in very large chokes that produce a large voltage drop (or that are too large physically).

4–1.4.1 *Crystal Oscillator Design Example*

Assume that the circuit of Fig. 4–14 is to provide an output at 50 MHz. The circuit is to be tuned by L_1. A 30V supply is available. The crystal will not be damaged by 30V, and will operate at 50 MHz with the desired accuracy. The MOSFET has an output capacitance of 3 pF, and will operate without damage with 30V on the drain. The desired output power is 40 to 50 mW.

Design of the bias network is the same as described in Chap. 2, except that R_L is omitted. Thus, the drain is operated at 30V (ignoring the small drop across L_1 and L_3). The values of R_1, R_2 and R_S should be chosen to provide an I_D that will produce 40 to 50 mW with 30V at the drain. A 45 mW output divided by 0.3 is 150 mW. Thus, the input power (and total dissipation) is 150 mW. Make certain that the MOSFET will permit a 150 mW dissipation at maximum anticipated temperature. For example, assume that the MOSFET has a 330 mW maximum dissipation at 25°C, a maximum temperature rating of 175°C, and a 2 mW/°C derating for temperatures above 25°C. If the MOSFET is operated at 100°C, or 75° above the 25°C level, the device must be derated by 150 mW (75 × 2 mW/°C), or 330 mW − 150 mW = 180 mW. Under these conditions, the 150 mW input power dissipation is safe.

With 30V at the drain, and a desired 150 mW input power, the I_D must be 150 mW/30V = 5 mA.

With a 30V supply, the output signal should be about 24V (30 × 0.8 = 24). Of course, this is dependent on the bias-feedback relationship.

To start, make C_2 three times the value of C_1 (plus the MOSFET output capacitance). With this ratio, the feedback signal will be 25 percent of the output, or 6V (24 × 0.25 = 6). Considering the amount of fixed and variable bias supplied by the bias network, a feedback of 6V may be large. However, the 6V values should serve as a good starting point.

For realistic values of L and C in the resonant circuit, let $C_1 = 2$ pF/meter, or 12 pF (50 MHz = 6 meters; 300/50 = 6).

With C_1 at 12 pF, and the MOSFET output capacitance at 3 pF, the value of C_2 is 45 pF (12 + 3 = 15; 15 × 3 = 45).

The total capacitance across L_1 is:

$$\frac{1}{\dfrac{1}{15} + \dfrac{1}{45}} \approx 12 \text{ pF}$$

With a value of 12 pF across L_1, the value of L_1 for resonance at 50 MHz is:

$$L \text{ (in } \mu\text{H)} = \frac{2.53 \times 10^4}{(50)^2 \times 12} \approx 0.84 \ \mu\text{H}$$

For conveninece, L_1 should be tunable from about 0.5 to 1.5 μH.

Keep in mind that an incorrect bias-feedback relation will result in distortion of the waveform, or low power, or both. The final test of correct operating point is a good waveform at the operating frequency, together with frequency stability at the desired output power.

The values of C_3 and C_4 should be 1/6.28 × (50 × 10⁶) × 5, or 630 pF. A slightly larger value (say, 1000 pF) will assure a reactance of less than 5 at the operating frequency.

The values of L_2 and L_3 should be 2000/6.28 × (50 × 10⁶), or 6.3 μH nominal. Any value between about 3 and 9 μH should be satisfactory. The best test for correct value of an RF choke in an oscillator is to check for RF at the power supply side of the line, with the oscillator operating. There should be no RF, or the RF should be a fraction of 1V (usually less than a few microvolts for a typical MOSFET oscillator). If RF is removed from the power supply line, the choke reactance is sufficiently high. Next, check for dc voltage drop across the choke. The drop should be a fraction of 1V (also in the microvolt range).

4-1.5 MOSFET Phase-Shift Oscillator

Figure 4–15 is the working schematic of an RC (resistance-capacitance) phase-shift oscillator. MOSFETs are well suited to RC circuits since no coupling capacitors are needed between stages (the MOSFET gate acts as a capacitor). Such oscillators are used at audio frequencies instead of the LC (inductance-capacitance) oscillators described in Sec. 4–1.4. RC oscillators avoid the use of inductances, which are not practical in the audio-frequency range. They are usually operated in class *A*, thus producing good waveforms.

The feedback principle is also used in RC oscillators. In the circuit of Fig. 4–15, the output (drain) of Q_1 is fed through three RC networks back to the gate of Q_1. Each network shifts the phase about 60°, resulting in an approximate 180° shift between drain and gate. Since the drain is normally shifted 180° from the gate, the RC shift of 180° brings the feedback to 360°, or back in-phase to produce oscillation. MOSFET Q_2 is used as an output amplifier.

bias network—Since both MOSFETs are operated at zero gate voltage, the Q-point drain voltage is set by I_{DSS}, and the values of R_4 and R_5. Both R_4 and R_5 are made variable. Resistor R_4 is adjusted to produce oscillations with good waveforms. Resistor R_5 is adjusted to produce the desired output swing. Typically, both Q_1 and Q_2 should be operated at one-half the supply voltage. For example, if I_{DSS} is 1 mA, and the supply is 30V, both R_4 and R_5 should be 15K, dropping both drains to about 15V.

output frequency—The oscillator frequency is determined by the RC time constant. To simplify design, the same values can be used in all three RC networks. However, such an arrangement will create a problem of power loss. Each of the RC networks functions as a low-pass filter. If the same values are used in all three sections, the signal loss will be about 15 dB through the networks. This loss, combined with the normal loss, could be sufficient to prevent oscillation if the voltage gain of Q_1 is low. The loop gain of any oscillator must be at least 1 (or slightly more for practical design). If the gain of Q_1 is 10 and the loss is anything greater than about 8 to 8.5, the circuit will not oscillate.

The RC network loss problem can be minimized by making the impedance of the succeeding network *greater* than that of the prior network. That is, there should be an impedance step-up as the signal passes from the drain of Q_1 to the gate of Q_2. For example, R_2 should be three times that of R_1; R_3 should be three times that of R_2. Thus, each RC network places very little load on the previous section and keeps loss at a minimum.

There should also be an impedance step-up between the output of Q_1

(set by R_4) and the first RC network. As a first trial, R_1 should be at least 3 times the value of R_4.

The values of C_1, C_2 and C_3 must be selected to produce the desired operating frequency. The output frequency is about equal to 1/(3RC). A more exact frequency calculation cannot be made in practical design, since MOSFET capacitance and resistance values must be added to the RC networks. However, the 1/(3RC) relationship is satisfactory for trial values.

The circuit of Fig. 4–15 is best suited to fixed-frequency use. It is diffi-cult to find a three-section variable capacitor that will track properly.

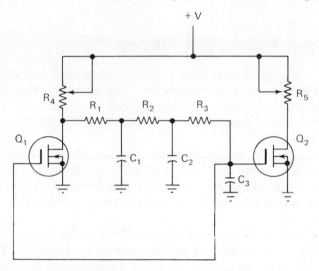

Fig. 4–15 MOSFET RC phase-shift oscillator.

4–1.5.1 *RC Phase-Shift Oscillator Design Example*

Assume that the circuit of Fig. 4–15 is to provide an output at 3.7 kHz. The power supply is 30V. The output signal is to be the maxi-mum possible without distortion. The MOSFETs have a zero gate voltage I_D of 1 mA.

For maximum output voltage swing, the drains of both Q_1 and Q_2 should be at one-half the supply or 15V.

With 1 mA of I_D flowing, the drops across R_4 and R_5 should be 15V. Thus, R_4 and R_5 should be 15K (15V/0.001 = 15K).

With R_4 at 15K, R_1 should be at least 45K. With R_1 at 45K, and a 3.7 kHz operating frequency, C_1 should be approximately 0.002 μF[1/(3 × 45K × 3.7 kHz] = 0.002 μF).

With R_1 at 45K, R_2 should be 135K (to get the impedance step-up). With R_2 at 135K, and a 3.7 kHz operating frequency, C_2 should be approximately 7000 pF [$1/(3 \times 135K \times 3.7$ kHz) $= 7000$ pF].

With R_2 at 135K, R_3 should be 405K. With R_3 at 405K, and a 3.7 kHz operating frequency, C_3 should be approximately 2200 pF [$1/(3 \times 405K \times 3.7$ kHz) $= 2200$ pF].

In practice, after the values have been selected and the components assembled, the gate of Q_2 is monitored on an oscilloscope, R_4 is adjusted for maximum signal swing without distortion, and R_5 is adjusted for maximum output swing without distortion, at the drain of Q_2.

4–2. CHOPPERS AND SWITCHES

MOS devices make excellent switches and choppers for many applications such as modulators, demodulators, sample-and-hold systems, mixing, multiplexing or gating, and similar circuits.

Several important characteristics of the MOSFET make it almost ideal for these applications.

One advantage is that there is *no inherent offset voltage* associated with MOSFETs as there is with two-junction transistors, since the conduction path between drain and source is predominantly resistive. That is, the conduction channel is either depleted or enhanced by controlling an induced field.

The dc gate input impedance of a MOSFET is also extremely high, and requires little power, since the gate is essentially a capacitor. The impedance is determined by the properties of the insulation layer of the gate.

Another important advantage of MOS devices used as switches is the exceptionally high ratio of OFF resistance to ON resistance of the drain-source channel. The resistance can be as low as about 50 ohms in the ON condition, and higher than thousands of megohms in the OFF condition. However, zero ON resistance is not possible with MOS devices (which may be a disadvantage in some applications).

The main disadvantage of MOS devices is the capacitance between the gate and drain, and the gate and source. This capacitance feeds through part of the gate control voltage to the signal path, and is detrimental to high-frequency signal isolation, also imposing a limitation on response times.

Figure 4–16 summarizes the advantages (and limitations) of MOS devices used as choppers and switches, compared to two-junction transistors and electromechanical relays. Note that the MOS device is generally superior, except for the zero ON resistance, and the feedthrough capacitance.

IDEAL CHOPPER	MOS	Two-junction	Electromechanical relay
CHARACTERISTICS			
Infinite life	good	good	poor
Infinite	good	good	poor
frequency			
response			
Infinite OFF resistance	good	fair	good
Zero ON resistance	poor	fair	good
Zero driving power consumption	good	fair	fair
Zero offset voltage	good	poor	good
Zero feedthru between driving signal and signal being chopped	fair	fair	good
Small size	good	good	poor

Fig. 4–16 Comparison of available chopper devices with an ideal. (Courtesy RCA)

4–2.1 MOS Characteristics Applicable to Choppers and Switches

Figure 4–17 shows the ohmic region of a MOSFET expanded for both positive and negative values of V_{DS}. In the ohmic region, when the device is fully ON, there is a linear relationship between drain current I_D and drain-to-source voltage V_{DS}. The magnitude of this resistance can be changed by varying the gate-to-source voltage V_{GS}. It is in the ohmic region that the MOSFET is useful for both chopper and switch applications.

To stay in the ohmic region, the drain current must be kept within narrow limits. That is, a relatively large value of load resistance R_L is required.

A very low value of leakage current is important in chopper applications, since this leakage current appears in the output circuit and produces an error voltage. The $I_{D(OFF)}$ versus temperature for a 2N4532 *P*-channel MOSFET is shown in Fig. 4–18. Below 100°C the leakage is so low that accurate readings are more dependent on available equipment and meas-

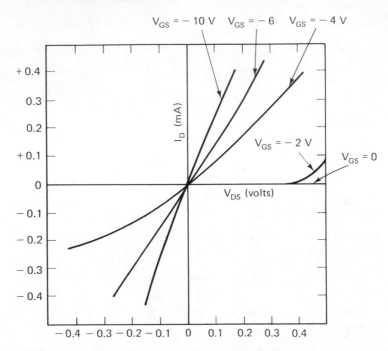

Fig. 4–17 2N4532 low-level (ohmic region) output characteristics. (Courtesy Motorola)

urement techniques than on the magnitude of the leakage current. However, the curve of Fig. 4–18 can be projected back to room temperature (shown with dotted lines) and an $I_{D(OFF)}$ of approximately 0.003 pA is read.

At room temperatures, surface and package leakage account for considerably more than I_{DSS} with a resulting room-temperature leakage of about 0.5 pA. This low leakage current indicates that the OFF voltage error (caused by leakage) will be negligible for most chopper and switch circuits. For an enhancement mode MOSFET, $I_{D(OFF)} = I_{DSS}$.

Drain-to-source resistance (R_{DS}), when the MOSFET is ON, is a very important characteristic in both chopper and switch circuits. Figure 4–19 illustrates R_{DS} versus V_{GS} for three values of temperature for a 2N4352.

On a static basis, there is interest in only two states of a MOS device — full ON or full OFF. The 2N4352 (a P-channel MOSFET) needs a negative potential of 10 to 20V to get an R_{DS} minimum. From Fig. 4–19, at $V_{GS} = -10$V and temperature 25°C, R_{DS} is about 300 ohms.

Compare this with Fig. 4–20 which shows the same characteristic for the MM2102 (an N-channel MOSFET). Note that for $V_{GS} = 10$V and temperature 25°C, R_{DS} is 100 ohms.

One reason for this 3:1 improvement in R_{DS} is that in the P-channel device, the carriers are holes, while in the N-channel MOSFET, the carriers are electrons. The mobility of electrons is greater than that of holes,

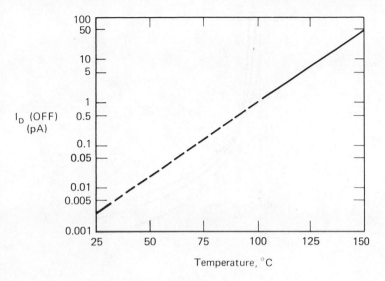

Fig. 4–18 2N4352 $I_{D(off)}$ versus temperature. (Courtesy Motorola)

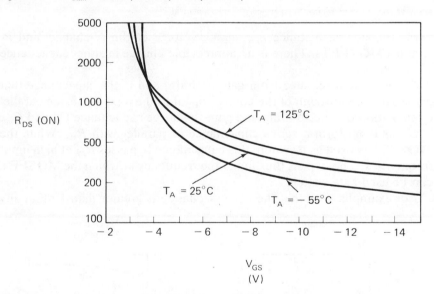

Fig. 4–19 2N4352 drain-source ON resistance. (Courtesy Motorola)

and thus is responsible for part of the improvement in R_{DS}. Since a low R_{DS} is needed in the ON condition, the N-channel MOSFET is preferable for choppers and switch applications.

It is very important to know how much capacitance *must* be charged and discharged during transition times. Figure 4–21 shows a plot of the small-signal, common-source, short-circuit input capacitance c_{iss}, and re-

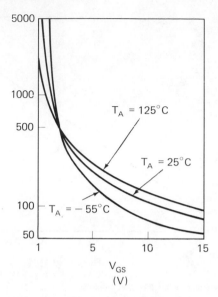

Fig. 4–20 MM2102 drain-source ON resistance. (Courtesy Motorola)

verse transfer capacitance c_{rss} versus voltage of the *N*-channel and *P*-channel MOSFETs. There is no appreciable change in either capacitance with voltage.

C_{rss} is the capacitance from gate-to-drain, and is the capacitance that causes the feedthrough of the control signal to the load. C_{iss} is the parallel combination of gate-to-drain and gate-to-source capacitance C_{gs} and C_{gd}.

C_{gs} and C_{gd} form a series capacitance in parallel with R_{DS}. When the MOSFET is used as a switch, this capacitance bypasses R_{DS} at high input frequencies. The bypass thus limits the frequency at which the MOSFET can be used as a switch.

For example, note that the c_{iss} capacitance is greater than 1 pF at any

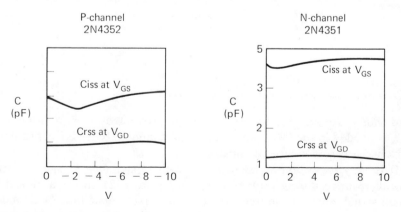

Fig. 4–21 C_{iss} and C_{rss} of two MOSFETs. (Courtesy Motorola)

V_{GS}. At a frequency of 100 MHz, the reactance of a 1 pF capacitor is about 1K. If the R_{DS} is greater than 1K, the 100 MHz signals will be bypassed around R_{DS}. If the capacitance is increased to 10 pF, the reactance drops to about 100 ohms. This will bypass most MOSFETs in the OFF condition.

temperature variations—The variation of MOSFET parameters with temperature can affect operation of a chopper circuit unless allowance is made for such variations in the circuit design. It is important to determine the approximate degree to which each parameter can be expected to change with temperature. Figure 4–22 shows curves of R_{DS}, I_{GSS}, and $I_{D(OFF)}$ as a function of temperature for a typical MOSFET.

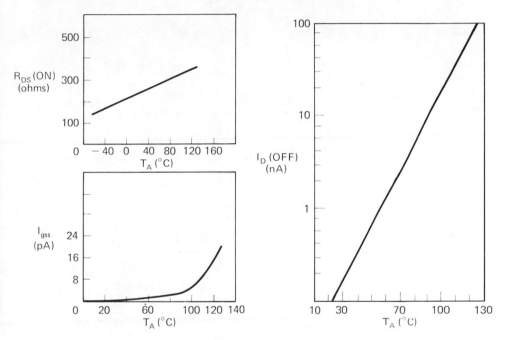

Fig. 4–22 Variation of 40460 parameters with ambient temperature. (Courtesy RCA)

4–2.2 Basic MOSFET Chopper Circuits

There are three basic chopper configurations: the series chopper, the shunt chopper, and the series-shunt chopper.

series chopper—The basic circuit, equivalent circuit, and equations for the series chopper are shown in Fig. 4–23. In order to operate in the ohmic region when the MOSFET is ON, the drain current must be limited to a low value. Ordinarily, a large value of load resistance is used to limit the current. The effects of load on chopper circuits is discussed in Sec. 4–2.3.

In the case of a series chopper, a high value of load resistance also minimizes the ON voltage error due to R_{DS}. The ON voltage error is given by:

$$E_{ON\ error} = \frac{E_S(R_S + R_{DS})}{R_S + R_{DS} + R_L}$$

When the MOSFET is OFF, a small error due to leakage of the MOSFET is present. This OFF error is given by:

$$E_{OFF\ error} = I_{DG}R_L$$

where

$$I_{DG} = I_{D(OFF)}.$$

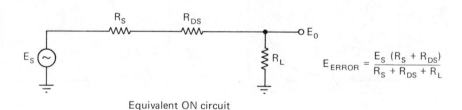

MOSFET series chopper

Equivalent ON circuit

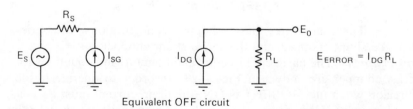

Equivalent OFF circuit

Fig. 4–23 MOSFET series chopper equivalent circuits. (Courtesy Motorola)

The MOSFET leakage is very low, and the resultant OFF voltage error is very small. Typically, with a MOSFET chopper with an R_L of 100K, the OFF error is less than one microvolt at room temperature.

shunt chopper—The simple shunt chopper shown in Fig. 4–24 performs the chopping function by periodically shorting the input to ground. From the ON equivalent circuit, Fig. 4–24(b), note that the shunt circuit is advantageous where a large source resistance R_S is present. The ON voltage error is given by:

$$E_{ON\ error} = \frac{E_S R_{DS}}{R_{DS} + R_S}$$

where R_L is much greater than R_{DS}.

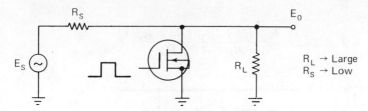

MOSFET shunt chopper

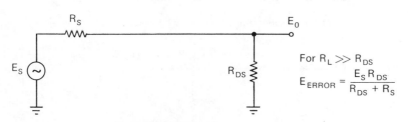

Equivalent ON circuit

For $R_L \gg R_{DS}$

$$E_{ERROR} = \frac{E_S R_{DS}}{R_{DS} + R_S}$$

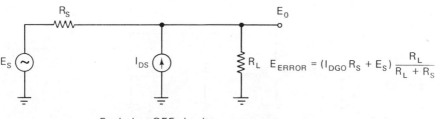

Equivalent OFF circuit

$$E_{ERROR} = (I_{DGO} R_S + E_S) \frac{R_L}{R_L + R_S}$$

Fig. 4–24 MOSFET shunt chopper equivalent circuits. (Courtesy Motorola)

When the MOSFET is OFF, the leakage current again produces an error voltage given by:

$$E_{OFF\ error} = \frac{I_{DGO}R_S R_L}{R_L + R_S}$$

However, the OFF error is usually small compared to the drop across R_S.

series-shunt chopper—The series-shunt chopper, Fig. 4–25, operates on the following principle: When Q_1 is ON, Q_2 is OFF. Conversely, when Q_1 is OFF, Q_2 is ON. The equivalent ON-OFF circuits are also shown in this figure. When Q_1 is ON and Q_2 is OFF, the output is similar to that of the series chopper, except for the small error introduced by the leakage of Q_2:

$$E_{ON\ error} = \frac{R_L\left[E_S + I_{DGO}\ (R_S + R_{DS})\right]}{R_L + R_S + R_{DS}}$$

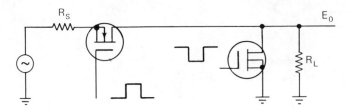

MOSFET series-shunt chopper

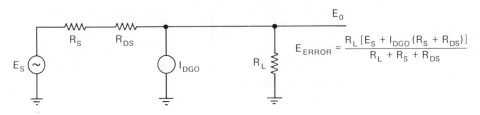

$$E_{ERROR} = \frac{R_L\left[E_S + I_{DGO}\ (R_S + R_{DS})\right]}{R_L + R_S + R_{DS}}$$

Equivalent ON circuit

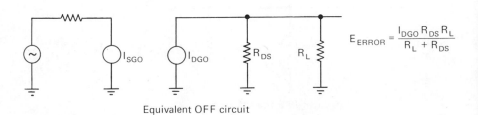

$$E_{ERROR} = \frac{I_{DGO}\ R_{DS}\ R_L}{R_L + R_{DS}}$$

Equivalent OFF circuit

Fig. 4–25 MOSFET series-shunt chopper equivalent circuit. (Courtesy Motorola)

When Q_1 is OFF and Q_2 is ON, the OFF voltage error due to the leakage of Q_1 is reduced since R_{DS2} appears in parallel with R_L. This can be seen from the OFF voltage error equation:

$$E_{OFF\ error} = \frac{I_{DGO}R_{DS}R_L}{R_L + R_{DS}}$$

However, the leakage current of a MOSFET is quite small, so the series-shunt circuit cannot be justified for the sake of minimizing the error due to leakage. The series-shunt circuit does have a definite advantage in the area of high-frequency chopping.

In the simple series chopper, when the MOSFET is OFF, c_{rss} must be discharged through the load resistor R_L. The relatively long time constant $(c_{rss}R_L)$ will limit the chopping frequency. In the series-shunt chopper, however, every time the series device turns OFF, the shunting device is turned ON, and the low resistance of Q_2 will parallel R_L. The RC time constant will, therefore, be greatly reduced, and the chopping frequency can be increased significantly.

4–2.3 Effect of Loads on MOSFET Choppers

Operation of all MOSFET chopper circuits is greatly affected by the magnitude of the source and load resistances, R_S and R_L. Figure 4–26 lists the output voltages of the three basic chopper circuits for various combinations of source and load resistances. It is assumed that the input voltage E_S is 1 mV, and the drain-to-source resistance R_{DS} is 100 ohms in the ON condition, and 1000 megohms in the OFF condition. The gate leakage resistance (typically 10^{12} ohms, or more) is neg-

Approximate output voltage E_0 (μV)
(Max. output 1 mV)

Source resistance	Load resistance	Shunt chopper		Series chopper		Series-shunt chopper	
R_S (ohms)	R_L (ohms)	ON	OFF	ON	OFF	ON	OFF
1 M	1 M	0.1	500	500	1	500	0.0001
100 K	1 M	1	900	900	1	900	0.0001
100	1 M	500	1000	1000	1	1000	0.0001
0	1 M	1000	1000	1000	1	1000	0.0001
1 M	100 K	0.1	90	90	0.1	90	0.0001
1 M	100	0.05	0.1	0.1	0.0001	0.1	0.00005
100 K	100 K	1	500	500	0.1	500	0.0001
100	100	333	500	333	0.0001	333	0.00005

Fig. 4–26 Steady-state chopper output voltage for various source and load resistance. (Courtesy RCA)

lected. The following conclusions can be drawn from the data shown in Fig. 4–26.

1. Only the series or series-shunt circuit should be used when R_S is less than $R_{DS(ON)}$.
2. In general, R_L should be high. In any event, R_L should be much greater than $R_{DS(ON)}$.
3. R_L should always be greater than R_S.
4. Performance of the series-shunt circuit is equal to or better than that of either the series or shunt chopper alone, for any combination of R_S and R_S.

4–2.4 Effect of Interelectrode Capacitances on MOSFET Choppers

The interelectrode capacitances of MOSFET choppers have their greatest effect as frequency increases. The capacitances are of little concern at low frequencies. The high-frequency effect of the capacitances are shown in Fig. 4–27, which is the ac equivalent of a MOSFET shunt chopper.

The input capacitance C_{gs} increases the rise time of the gate driving signal, and thus the switching time of the chopper. This effect is not usually a serious limitation, however, because the switching time of the MOSFET depends primarily on the input and output time constants. Switching times as short as 10 nS can be achieved when a MOSFET is driven from a low-impedance source, and the load resistance is less than about 2000 ohms.

The output capacitance C_{ds} also tends to limit the maximum frequency that can be chopped. When the reactance of this capacitance becomes much lower than the load resistance R_L, the chopper becomes ineffective because XC_{ds} is essentially in parallel with R_L and $R_{DS(OFF)}$.

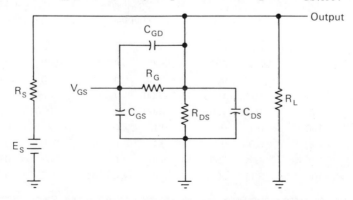

Fig. 4–27 AC equivalent circuit of MOSFET shunt chopper. (Courtesy RCA)

The feedthrough capacitance C_{gd} is the most important of the three interelectrode capacitances because it couples a portion of the gate drive signal into the load circuit, and causes a *voltage spike* to appear across R_L each time the gate drive signal changes stage. C_{gd} and R_L form a differentiating network that allows the leading edge of the gate drive signal to pass through. The output capacitance C_{ds} is beneficial to the extent that it helps reduce the amplitude of the feedthrough spike.

The effect of the feedthrough spikes can be reduced by several methods. Typical approaches include the following:

1. Use of a clipping network on the output when the input signal to be chopped is fixed in amplitude.
2. Use of a low chopping frequency.
3. Use of a MOSFET that has a low feedthrough capacitance (typically a fraction of one pF).
4. Use of a gate drive signal that has poor rise and fall times (sloppy square wave).
5. Use a source and load resistance as low as feasible. Of course, low values of R_S and R_L produce greater error voltages, as discussed in Sec. 4-2.3.
6. Use of shield between the gate and drain leads.
7. Use of a series-shunt chopper circuit.

4-2.5 Practical MOSFET Chopper Circuits

In the following paragraphs, some actual chopping circuits are examined to determine the limitations of input voltage and chopping frequency.

practical series chopper—Figure 4-28 shows a simple, but complete, series chopper circuit. The maximum chopping frequency is about 200 kHz, depending on the capacitances of the MOSFET. This limitation is primarily due to the long RC time constant for discharging c_{rss} through the 10K resistor. The time constant can be shortened, and the frequency increased, if the 10K resistor is reduced. However, this results in a greater error voltage.

The maximum allowable input voltages are set by the V_{GS} limits of the MOSFET. In addition, there is a minimum input voltage limitation caused by feedthrough spikes in the output channel. This feedthrough is caused by c_{rss} and c_{iss}, as previously discussed. For inputs less than about 10 mV, the feedthrough spikes become an appreciable part of the output waveform (particularly at high frequencies).

As already mentioned, there are several circuit techniques useful for minimizing these spikes. First, the control signal (pulse generator) at the gate can be a sloppy square wave. That is, the dv/dt of the input pulse should be kept as low a value as possible. Sharp corners on the waveform

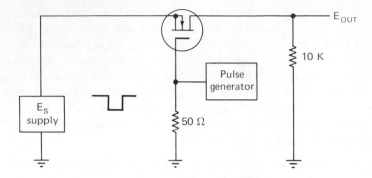

Maximum chopping frequency F(Max) ≈ 200 kHz
Maximum input voltage E_S (Max) ≈ + 2 V − 0.4 V

Fig. 4–28 Practical series chopper using *N*-channel MOSFET.

should be avoided. (A sine wave could be used.) Next, a capacitor can be connected across the output to filter the spikes. Finally, if a fixed amplitude output is acceptable, a clipper circuit can be connected across the output.

practical series-shunt chopper for high-frequency use—Figure 4–29 is a series-shunt high-frequency chopper using *complementary* enhancement-mode MOSFETs.With the components shown, the circuit will operate satisfactorily at frequencies up to about 5 MHz. An *N*-channel and a *P*-channel MOSFET are used as the series and shunting devices, respectively. This allows one drive circuit for both devices.

When the series MOSFET is OFF, the R_{DS} of the shunting MOSFET

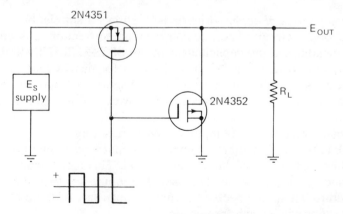

Maximum chopping frequency F(Max) ≈ 5 MHz
Maximum input voltage E_S (Max) ≈ + 0.5 V, − 4 V

Fig. 4–29 Series-shunt chopper for high-frequency applications using complementary enhancement mode MOS-
FETs. (Courtesy Motorola)

is about 200 ohms. This value parallels R_L to ground, and reduces the net output load resistance to about 200 ohms (the parallel combination of 200 ohms and 10 K). Thus, the RC time constant is reduced to 2 percent of its original value.

This circuit can also be modified to accept large values of input voltage. Such a circuit is described in Sec. 4–2.6.

practical series-shunt chopper for low input voltages—A series-shunt chopper capable of low-level chopping is shown in Fig. 4–30. Two *N*-channel MOSFETs *with matched* c_{rss} are used in the circuit. The gate drives for the pair are produced by a current-mode astable multivibrator. A current-mode multivibrator is one where high switching currents are used, resulting in good frequency stability. The main reason for using a current-mode multivibrator here is so that the complementary outputs of the MOSFET are not delayed in time, with respect to each other. That

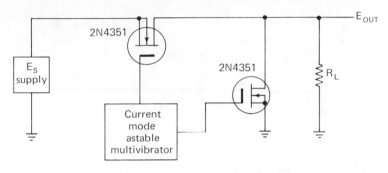

Maximum chopping frequency $F(Max) \approx 5$ MHz
Maximum input voltage E_S (Min) $\approx \pm 10 \mu V$

Fig. 4–30 Series-shunt chopper for low input voltages. (Courtesy Motorola)

is, when one output is turning OFF, the other output must be turning ON.

By matching c_{rss}, the feedthrough spikes in the output of the chopper can be nearly eliminated. Complete elimination of feedthrough spikes is difficult to obtain since the turn-on and turn-off characteristics of MOS-FETs are not symmetrical.

dual-gate MOSFET chopper circuits—The circuits shown in Figs. 4–31 and 4–32 use dual-gate MOSFETs in chopper or gating circuits. In the shunt choppers of Fig. 4–31, the MOSFET is normally conductive, therefore, e_o is low. A negative gating pulse turns off the MOSFET so that approximately 50 percent of e_g appears at the output terminals. The circuit of Fig. 4–31(a) features the use of an additional control potential V_{G2}. A dc potential may be applied to the second gate, establishing the value of desired channel ON resistance, or $R_{DS(ON)}$. As an alternative, the second gate can function as a *coincidence-gate* to reduce e_o to a low value.

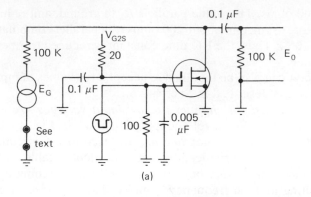

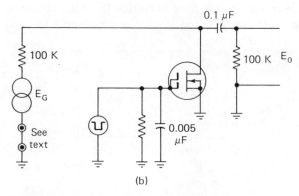

Fig. 4–31 Dual-gate shunt chopper. (Courtesy RCA)

This requires a positive-going pulse applied to gate 2 simultaneously with a positive-pulse to gate 1.

The circuits illustrated in Fig. 4–32 function in a manner opposite to those of Fig. 4–31. That is, output voltage appears at e_o in the absence of a gating signal. Consequently, a negative gating signal reduces the level of e_o. The dual-gate configuration can be made into an OR circuit. That is, a negative signal applied to gate 2 of sufficient magnitude to override V_{G2} will also reduce the level at e_o.

The circuits of these two figures show a jumper connected between two terminals in the drain-to-ground-return circuits. The circuits assume a peak generator level e_g of less than 0.2V. Should the signal exceed this value, it is possible that the parasitic *diode* between the drain and semi-conductor substrate will be driven into conduction and load the signal. This can be overcome by connecting a suitable dc potential in lieu of the jumper, so that a positive potential is applied to the drain. The magnitude of this voltage should equal or exceed the peak value of the signal from e_g.

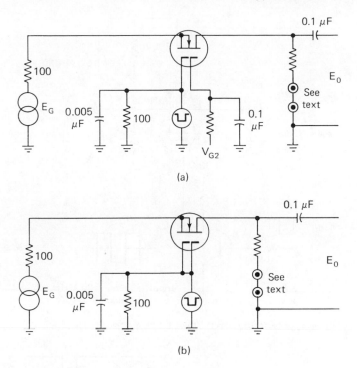

Fig. 4–32 Dual-gate series chopper. (Courtesy RCA)

4–2.6 MOSFET Analog Switching Circuits

Most of the information described thus far for MOSFET choppers can be applied to MOSFET analog switches. By definition, the analog switch is a device that either transmits an analog signal without distortion, or completely blocks it off.

A MOSFET analog switch is shown in Fig. 4–33. Of course, the circuit can also be used as a chopper. There may be a problem with the circuit when large plus or minus values of input voltage are present. Although there is no PN junction at the gate, there are PN junctions from substrate-to-source and substrate-to-drain. These junctions must not be allowed to become forward biased. One way to accomplish this is to cut off the substrate lead, and leave it floating. However, since the substrate is connected to the can, this results in quite a bit of pick-up. A more practical solution is to couple the junctions with diodes. For full protection, three diodes should be connected as shown.

commutator using MOSFET analog switches—The analog switch can be used in a commutator circuit, as shown in Fig. 4–34. Each switch

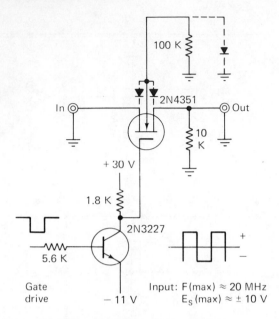

Fig. 4–33 MOSFET analog switch for large input voltages. (Courtesy Motorola)

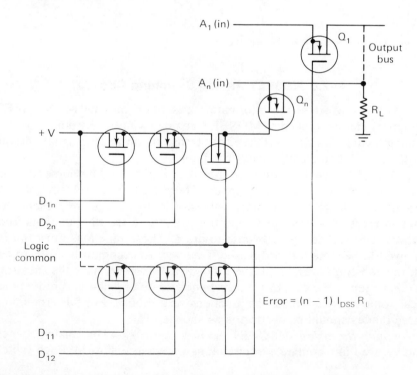

Fig. 4–34 Commutator network using *N*-channel MOSFETs. (Courtesy Motorola)

has a three-input AND gate in series with the gate drive. In order to turn a switch ON, a positive potential is required at the gate. To accomplish this, the three inputs of the AND gate must be "true" (at a positive potential).

Assume that A_n is to be sampled. Suppose the logic common is a clock signal, that D_{1n} is an order from the control system to sample A_n, and D_{2n} is a ready signal from the device to be sampled. When all of these conditions are true at the same time, switch Q_n is turned ON, and A_n is sampled. The circuit of Fig. 4–34 can be modified to accept large values of input voltage.

In this type of commutating circuitry, where only one channel is turned ON at a time, an error is introduced due to the leakages of the other FET switches. Assuming that the leakage is the same for all switches, an approximation of the error signal is given by $(n - 1) \times I_{DSS} \times R_L$.

4–3. ATTENUATORS AND VOLTAGE-VARIABLE RESISTORS

When a MOSFET is used in the ohmic region below V_P (Fig. 4–17), the MOSFET is simply a voltage-controlled resistor (or voltage-variable resistor, VVR). The uses of a MOSFET in the ohmic region are virtually unlimited. A classic example is the use of a MOSFET as an attenuator. In effect, a MOSFET can be used anywhere a small-signal voltage-controlled resistor is needed. Of course, heavy currents cannot be handled by present-day MOSFETs.

Of prime interest is the curve of Fig. 4–35, where the R_{DS} (normalized

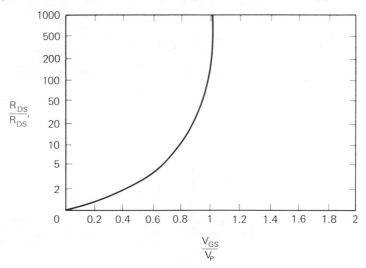

Fig. 4–35 Normalized R_{DS} data for depletion silicon MOSFETs. (Courtesy Motorola)

to R'_{DS} or the drain-source resistance at zero bias) is plotted against a ratio of V_{GS}/V_P. That is, the drain-source resistance is given as a function of the gate-source voltage V_{GS} normalized to V_P. Here V_P is defined as the gate-source voltage required to reduce I_D to 0.001 I_{DSS}. Note that when V_{GS} is equal to V_P, the drain-source resistance increases rapidly (to infinity, in theory). This is valid for several families of depletion MOSFETs. However, enhancement MOSFETs are excluded.

4–3.1 Dynamic Attenuators

Figure 4–36 shows a dual-gate MOSFET in an attenuator

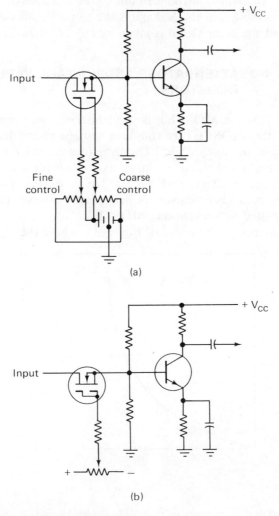

(a)

(b)

Fig. 4–36 MOSFET dynamic attenuators. (Courtesy RCA)

circuit. In Fig. 4–36(a) both gates are used as control elements. This type of circuit is particularly attractive when control of the attenuator must be located at some remote point. A dc potential on gate 1 has greater control on the channel resistance than is the case for gate 2. Thus, an arrangement can be used whereby gate 2 provides a *fine* attenuator adjustment, and gate 1 controls the course adjustment.

Figure 4–36(b) depicts the dual-gate MOSFET used as a triode-connected attenuator circuit. Curves showing typical variations in channel resistance as a function of gate-voltage are given in Fig. 4–37.

Figure 4–38 shows two MOSFETs operated as voltage-variable resistors to form a dynamic attenuator. The curves of this figure show the ability of a change in V_{GS} to R_{DS}. Here a change of only 1 V swings the attenuation over 50 percent of the range. As the resistance of one MOSFET increases, that of the other decreases. If the drain-source voltage is permitted to exceed V_P, as it has in the case of $V_{IN} = 8$V, it appears that R_{DS} still varies. (Here the series element is acting as a controlled-current source to the load.)

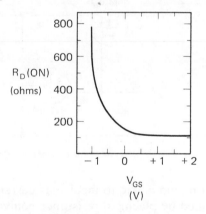

Fig. 4–37 ON resistance as a function of gate voltage for dual-gate MOSFET used as dynamic attenuators. (Courtesy RCA)

automatic gain control (AGC)—A MOSFET operated as a voltage-controlled resistor can be combined with other active devices to form a simple AGC circuit, such as shown in Fig. 4–39. Here the MOSFET changes the voltage gain over a 30-dB range with only a 1 V change in V_{GS}. This circuit has some harmonic distortion, due to the unbypassed emitter degeneration.

When the MOSFET is used in conjunction with an operational amplifier, as in Fig. 4–40, the gain can be made variable by changing the ratio of R_F/R_{IN}. This is also a rather simple AGC circuit, but with improved distortion characteristics.

Note that a MOSFET is used across the R_{IN} resistor, and across the offset-minimizing resistor (at the + input of the op-amp). Offset voltage

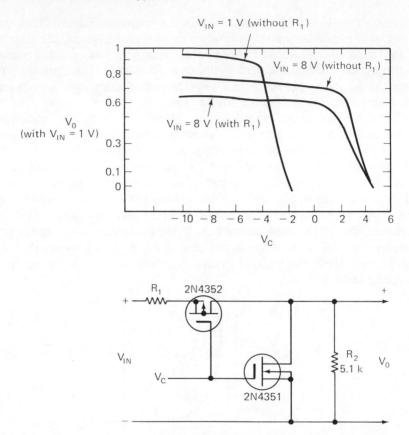

Fig. 4–38 Dynamic attenuator using two MOSFETs as voltage-variable resistors. (Courtesy Motorola)

at the output of an op-amp is due to input bias currents. These currents are normally minimized by placing a resistance equivalent to the parallel combination of R_F and R_{IN} at the noninverting $(+)$ input of the op-amp. If a MOSFET is used only across R_{IN}, the input impedance will change with variations in control voltage, shifting (or offsetting) the output voltage. The MOSFET across the minimizing resistor permits the resistance to vary according to the variation in R_{IN}. In turn, this minimizes the change in output offset voltage due to the changing R_{IN}. The value of R_{IN} should equal R_F to insure that gain will not drop below unity.

4–4. MOSFET CURRENT REGULATORS

MOSFETs are well suited to circuits requiring a constant current (or current regulation). Output impedances of MOSFETs can range anywhere from the kilohm region to the tens of megohms, de-

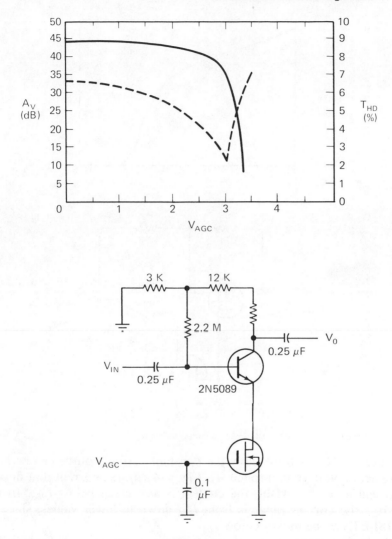

Fig. 4–39 MOSFET operating as a voltage-controlled resistor in an AGC circuit. (Courtesy Motorola)

pending on the device and configuration used. MOSFETs are available with values of g_{os} (output conductance) ranging from one-half to several hundred micromhos. Saturation voltages, or the minimum operating voltages, are in the region of 1V. The maximum voltage before breakdown approaches the 100V level. A nearly zero temperature coefficient is practical if the MOSFET is biased properly.

Most importantly, the MOSFET constant-current source is a very simple, easy-to-design circuit. Either a fixed or an adjustable current source can be constructed with nothing more than a MOSFET, a resistor (fixed or variable), and a power source.

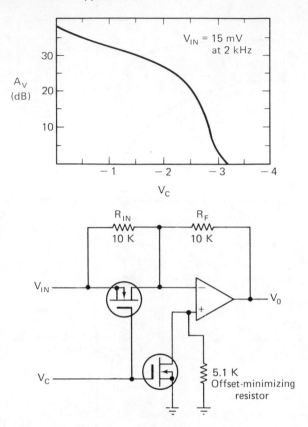

Fig. 4–40 IC amplifier with MOSFET AGC circuits. (Courtesy Motorola)

Figure 4–41 illustrates how the constant-current source can be fixed, variable, or voltage-controlled. In Fig. 4–41(a), the current that flows is I_{DSS}, and in Fig. 4–41(b), the current is any value below I_{DSS}. In Fig. 4–41(c), the current range is from I_{DSS} down to lower values, since the MOSFET can be biased below I_{DSS}.

The basic circuits shown in Fig. 4–41 can be put in series for high voltages, provided the circuits are shunted with balancing resistors. The circuits can also be put in parallel for higher currents. The MOSFETs can be connected in series-opposing for bilateral current limiting.

4–4.1 Fixed Constant Current

The simplest constant-current circuit is that of Fig. 4–41(a). The circuit is essentially a MOSFET with the gate and source shorted. In this case, the MOSFET operates at I_{DSS}, the zero-bias drain current. The circuit output conductance g_o is equal to the g_{os} of the MOSFET. At low frequencies, g_{os} (the real part of y_{os}) is equal to y_{os}. For higher frequencies,

Fixed constant current

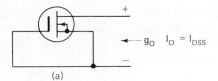

(a)

Variable constant current

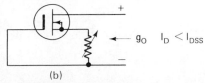

(b)

Voltage controlled current source

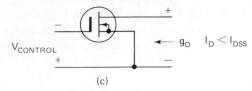

(c)

Fig. 4–41 Basic MOSFET constant-current regulators. (Courtesy Motorola)

depending on the value of the MOSFET output capacitance, it may be necessary to add a series inductance to offset the change in y_{os} (or the difference between g_{os} and y_{os}).

The output current I_O of the circuit changes if V_{DS} changes, according to the relationship:

$$\Delta I_O = \Delta V_{DS} g_{os}$$

For example, assume that $g_{os} = y_{os} = 250\ \mu\text{mhos}$, and there is a 1V change in V_{DS}. This will produce a 250 μA change in I_O.

The circuit of Fig. 4–41(a) will deliver a relatively constant current from about $2V_P$ (the pinch-off or threshold voltage) to BV_{DS} (the breakdown voltage drain-to-source). This range is illustrated in Fig. 4–42.

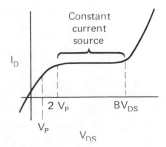

Fig. 4–42 MOSFET characteristics as constant-current regulator. (Courtesy Motorola)

4–4.2 Adjustable Constant Current

With the addition of a source resistor, as shown in Fig. 4–41(b), the circuit becomes capable of supplying any current below I_{DSS}. The approximate value of gate-source voltage V_{GS} required for a given operating current is:

$$V_{GS} = V_P\left(1 - \sqrt{\frac{I_O}{I_{DSS}}}\right)$$

With V_{GS} established, the value of source resistance R_S is:

$$R_S = \frac{V_{GS}}{I_O}$$

Resistor R_S can be varied to provide an adjustable current source. As R_S is increased and I_O decreased, the MOSFET g_{os} decreases. The circuit output conductance g_o decreases more rapidly than the MOSFET g_{os} because of the feedback action produced across R_S. The circuit output conductance is:

$$g_o = \frac{g_{os}}{1 + R_S(g_{os} + g_{fs})} \approx \frac{g_{os}}{1 + R_S g_{fs}}$$

where g_{fs} is the real part of y_{fs}, the forward transfer admittance.

4–4.3 Voltage-Controlled Constant Current

The constant-current circuit can be controlled by an external voltage, as shown in Fig. 4–41(c). The approximate value of control voltage ($V_{control}$) required for a given I_0 is:

$$V_{control} \approx V_P\left(1 - \sqrt{\frac{I_O}{I_{DSS}}}\right)$$

For example, assume that V_P is 3V, I_{DSS} is 1 mA, and the desired I_0 is 0.5 mA (I_0 must be less than I_{DSS}). Then:

$$V_{control} \approx 3\left(1 - \sqrt{\frac{0.5}{1}}\right) \approx 0.9V$$

Figure 4–43 illustrates a dual-gate MOSFET constant-current source, where one gate is used to provide the voltage control. The use of a dual-gate MOSFET provides higher values of current regulation.

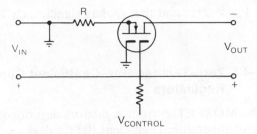

Fig. 4–43 Dual-gate MOSFET as constant-current source.

4–4.4 Cascaded MOSFET Current Regulators

If two MOSFETs are cascaded as shown in Fig. 4–44, a much lower output conductance g_o value (for a given I_0) can be obtained. Here I_0 is regulated by Q_1. Note that V_{DS} of Q_1 is equal to V_{GS} of Q_2. The dc value of I_0 is controlled by R_S and Q_1. However, Q_1 and Q_2 both affect current stability. Where $R_S = 0$, then:

$$I_0 = \frac{V_{DS2}g_{os1}g_{os2}}{g_{os1} + g_{fs2}}$$

and

$$g_o = \frac{g_{os1}g_{os2}}{g_{os1} + g_{os2} + g_{fs2}}$$

if $R_S = 0$ and $g_{os1} \approx G_{OS2}$

$$g_o \approx \frac{(g_{os})^2}{g_{fs}\,(1 + R_S g_{fs})}$$

When designing cascaded MOSFET current sources, use care to ensure that both MOSFETs are operating with adequate drain-source volt-

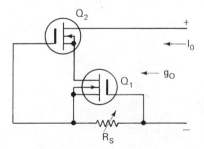

Fig. 4–44 Cascaded MOSFET current regulators. (Courtesy Motorola)

age, preferably $V_{DS} > 2V_P$, and that Q_2 has significantly higher I_{DSS} than Q_1.

4–4.5 Zero-Temperature-Coefficient Current Regulators

The MOSFET current regulators described thus far can be operated as zero-temperature-coefficient ($0TC$) devices. Of course, the MOSFET must be operated at a specific current I_{DZ} as discussed in Sec. 2–2 of Chap. 2.

The approximate value of $0TC$ current I_{DZ} is found by:

$$I_{DZ} \approx I_{DSS} \left(\frac{0.63}{V_P}\right)^2$$

The gate-source bias voltage for $0TC$ is:

$$V_{GSZ} \approx V_P - 0.63$$

Note that I_{DZ} increases as I_{DSS} increases. Typically, I_{DZ} can be as high as 1 mA for I_{DSS} units of 20 mA.

By operating the I_D below, but near, I_{DZ}, the temperature coefficient is positive. Conversely, negative temperature coefficients will result if $I_D > I_{DZ}$.

4–4.6 Current Regulators as Voltage References

Low-voltage references can be made by using a MOSFET current source in series with a resistor, as shown in Fig. 4–45. The source drives a resistor of known value, producing an output reference voltage, which can be determined by Ohm's law. For example, if I_{DSS} is 1 mA and R is 1K, the voltage reference is 1V.

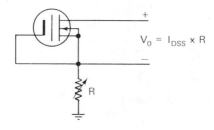

Fig. 4–45 MOSFET current regulator as voltage reference. (Courtesy Motorola)

Figure 4–46 illustrates a dual-gate MOSFET used as a variable-voltage reference. The output reference voltage is determined by the value of R, and the I_D. In turn, the I_D is set by the voltage at gate 2.

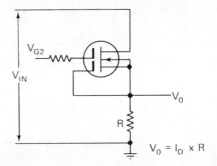

Fig. 4–46 Dual-gate MOSFET as variable voltage reference.

4–4.7 MOSFET Current Regulators as Zener Source Elements

A MOSFET current regulator can be used in place of the conventional series resistor as a Zener source element. Such an arrangement has several advantages. For example, the maximum permissible V_{IN} is determined by the maximum breakdown voltage of the MOSFET current regulator, rather than the maximum Zener dissipation. Thus, a low-voltage Zener can be used in a high-voltage system (provided the MOSFET can stand the high voltage). Variations in V_{IN} have little or no effect on V_{OUT}. Thus, power efficiency is increased.

Figure 4–47 shows the comparison between a fixed resistor and a MOSFET as a Zener source element.

4–4.8 Zero-Temperature-Coefficient MOSFET/Zener Combinations

Combining the $0TC$ feature of a MOSFET regulator with a $0TC$ Zener diode provides a highly temperature-stable, yet simple, reference voltage. This requires a circuit similar to that of Fig. 4–41(b) for the MOSFET. However, the source resistor value must be such that the resultant V_{GS} is V_{GSZ}. Refer to Secs. 2–2 and 4–4.5.

Figures 4–48 and 4–49 depict the basic MOSFET/Zener circuit for $0TC$ operation, as well as the advantages for such an arrangement. Note that the Zener diode *requires a current identical* to the I_{DZ} of the selected MOSFET.

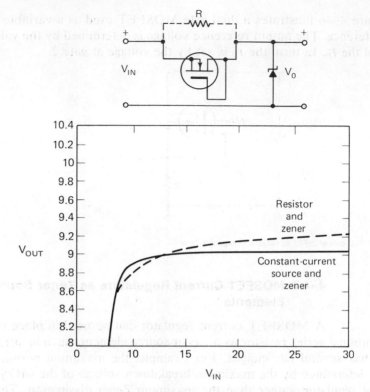

Fig. 4-47 Comparison between a fixed resistor and a MOSFET as Zener source elements. (Courtesy Motorola)

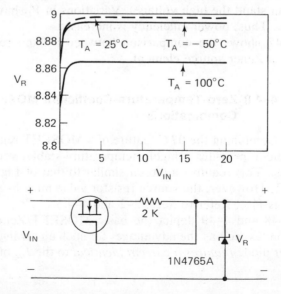

Fig. 4-48 MOSFET current regulator as 0TC voltage reference. (Courtesy Motorola)

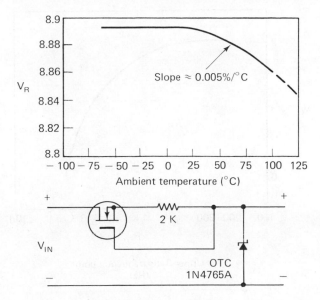

Fig. 4–49 MOSFET current regulator as 0*TC* voltage reference across a given temperature range. (Courtesy Motorola)

4–5. MOSFET FILTERS

When a MOSFET operated as a voltage-controlled resistor (Sec. 4–3) is combined with passive components such as capacitors or inductors, voltage-controlled frequency tuning can be implemented. Figure 4–50 shows a voltage-controlled low-pass filter. Here, the change in V_{GS} of 1V alters the upper 3-dB frequency point by a decade.

A high-pass filter can be implemented by transposing the MOSFETs and capacitors (capacitors in series with the line, and MOSFETs across the line). Likewise, high-pass and low-pass filters can be combined to form band-pass filters. Refer to the author's *Handbook of Simplified Solid-State Circuit Design,* Prentice-Hall, Inc., Englewood Cliffs, New Jersey, 1971, for a detailed discussion of RC filters. The following is a summary.

Filter attenuation is rated in terms of dB drop at a given frequency. Generally, RC filters are designed to produce a 3-dB drop (to 0.707) of input at a selected cutoff frequency. The relationships for capacitance and resistance values versus cutoff frequency for RC filters with a 3-dB drop are:

$$\text{cutoff frequency} \approx \frac{1}{6.28RC}$$

where C is in farads, R is in ohms, and cutoff frequency is in Hz.

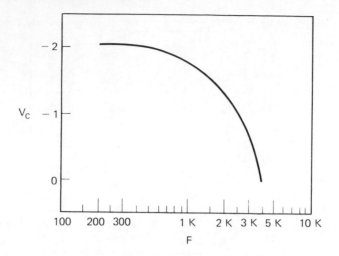

Upper 3-dB frequency point
(Hz)

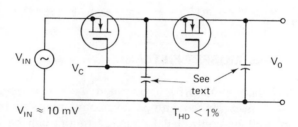

Fig. 4-50 MOSFET as voltage-controlled low-pass filter. (Courtesy Motorola)

A single RC filter will provide a gradual transition from the passband to the cutoff region. If a rapid transition is necessary for design, two or more RC filters can be combined. (Two low-pass stages are shown combined in Fig. 4–50.) As a rule of thumb, each stage will increase the attenuation by 6 dB at the cutoff frequency.

If the drain-source resistance for a given control voltage is known (or can be measured), it is a simple matter to calculate the required value of C to produce a 3 dB drop. For example, assume that the drain-source resistance is 1K at a given control voltage, and it is desired to produce a 3 dB drop at 300 Hz. The value of C is:

$$C \approx \frac{1}{6.28 \times 1000 \times 300} \approx 0.5 \ \mu F$$

The circuit of Fig. 4–51 is a *frequency-selective amplifier-type of filter* intended for operation within the audio range of 10 Hz to 20 kHz. The

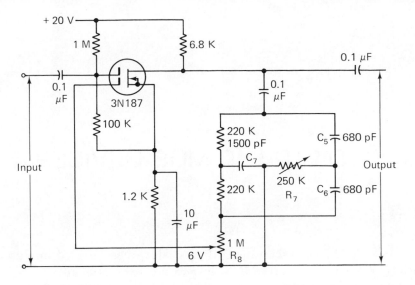

Capacitor values

Frequency (Hz)	C_5, C_6 (pF)	C_7 (pF)
150	5,600	12,000
300	2,700	6,200
600	1,300	3,000
2400	330	750
4800	160	360
9600	82	180

Fig. 4–51 Dual-gate MOSFET as frequency-selective amplifier type filter. (Courtesy RCA)

circuit requires a dual-gate MOSFET. Such frequency-selective filters are used for selective coding (such as garage-door openers), for narrowing the bandwidth response in CW receivers to eliminate unwanted sidebands, and in systems requiring some form of keying impulse (such as synchronizing the narration in a tape recorder with slides).

The circuit shown in Fig. 4–51 is an audio amplifier with a twin-T RC filter circuit in the output. This network provides regenerative feedback to the input circuit at an audio frequency predetermined by the selection of capacitors C_5, C_6 and C_7. The peaking control R_7 fine-tunes the twin-T for the desired frequency of operation, and potentiometer R_8 adjusts the level of feedback for desired performance.

The values shown for this circuit provide selection at a frequency of 1200 Hz. Figure 4–51 also lists circuit values for operation at other frequencies.

5. DIGITAL IC MOS DEVICES

   Unlike discrete MOSFETs which can be used in a great variety of applications, IC MOS devices are used primarily for digital logic circuits. To make full use of such digital MOS devices, the user *must be familiar with digital logic,* including simplification and manipulation of logic equations, working with logic maps, and implementing basic logic circuits (counters, registers, memories, etc). All of these subjects are discussed in the author's *Handbook of Logic Circuits,* Reston Publishing Company, Inc., Reston, Virginia, 1972.

 In this chapter, we shall concentrate on the types of digital IC MOS devices available, their relative merits, and the interpretation of manufacturer's data. Some typical applications for the digital MOS elements, as well as notes on how one type of digital MOS device can be made to work with other types of digital logic in a given system, are discussed. Also included are discussions concerning special applications for digital MOS devices, such as forming multivibrators, oscillators, etc.

 Before going into the details of digital IC MOS devices, let us consider the "what" and "why" of MOS in digital applications.

 MOS integrated circuits require only one-third of the process steps needed for the standard double-diffused two-junction IC. The most significant feature of MOS ICs is the large number of semiconductor circuit elements that can be put on a small chip. The size relationship of MOS and two-junction ICs is shown in Fig. 5–1. This high circuit density means large-scale integration (LSI), instead of medium-scale integration (MSI). For example, it is possible to put 5,000 devices on a silicon chip only 150×150-mils square. Each transistor in a typical MOS/LSI array requires as little as 1 square mil of chip area, a great reduction over the two-junction IC transistor which requires about 50 square mils.

 MOS/LSI has several advantages over two-junction MSI. These include:

lower cost per circuit function
fewer subsystems to test
fewer parts to assemble and inspect
increased circuit complexity per package
lower power drain per function
a choice of standard or custom products.

This last advantage is of particular importance to the user. Most major MOS device manufacturers offer a complete line of standard digital ICs. Typically, the line will include gates, switches, registers, dividers, counters, generators, synthesizers, memories, coders, decoders, and general purpose logic. The great majority of design problems can be solved with these standard, off-the-shelf devices.

In addition to the standard ICs, many MOS manufacturers offer a custom production service. That is, they will produce complete logic devices from the customer's logic drawings. In effect, the customer draws the de-

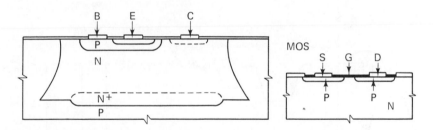

Fig. 5–1 Comparison of two-junction and MOSFET size. (Courtesy Texas Instruments)

sired logic, defines the inputs and outputs, and describes the test procedures. This information is then sent to various MOS manufacturers for bids on completed hardware. Some MOS suppliers also provide software for both standard and custom devices. Likewise, manufacturers will sometimes produce MOS logic elements from customer's software.

From this description, it would appear that the customer need only tell the manufacturer what is wanted, check the bids, and then wait for the finished hardware. However, it is necessary that the customer or user have a working knowledge of MOS logic devices in order to make an intelligent comparison of the manufacturer's services. For example, the user should know the switching characteristics of MOS devices, and the basics of complementary logic.

The purpose of this chapter is to acquaint the readers with MOS logic, in general, so that they can select commercial units to meet their particular circuit requirements. If the needs cannot be met with existing devices, the information in this chapter will provide a sound basis for venturing into the uncertain world of custom MOS logic hardware.

Keep one point in mind, no matter what equipment or device is involved, from the design standpoint, MOS/LSI is a two-dimensional layout, rather than three-dimensional. Mathematically, MOS/LSI operations can be predicted, and mathematic models lend themselves to computer-aided design analysis. Therefore, the circuit can be laid out and its operation checked before it is built.

5–1. MOS DEVICE SWITCHING CHARACTERISTICS

All solid-state logic circuits are based on the switching characteristics of the devices. The characteristics that affect logic circuit design can be examined by looking at the basic inverter circuit using a MOSFET. An inverter circuit, with an equivalent circuit for the MOSFET, is shown in Fig. 5–2.

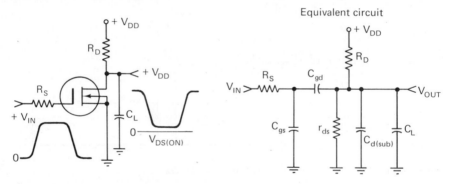

Fig. 5-2 Basic MOSFET inverter. (Courtesy Motorola)

The first characteristic of importance is that the MOSFET inverter circuit does not require an OFF bias. The input is represented simply as a source resistance, and an ideal pulse generator. With the input at zero, the MOSFET is OFF. That is, the output is at the high logic-level ($+V_{DD}$), and the only current flowing is I_{DSS}. Typically, MOS devices used in logic circuits have an I_{DSS} in the range of nanoamperes.

In logic work, the I_{GSS} of the MOSFET is of little interest. Typically, the gate leakage current is several orders of magnitude below that of a JFET, and even lower than the comparable characteristic of a two-junction device. Thus, the only remaining static characteristic of interest is $V_{DS(ON)}$.

With the input at $+V_{IN}$, the MOSFET is biased to the ON condition, and the output is at the low logic-level $V_{DS(ON)}$. This characteristic for a typical N-channel MOSFET is shown in Fig. 5–3. Note that the $V_{DS\,(ON)}$

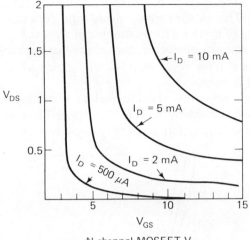

Fig. 5–3 ON drain-source voltage. (Courtesy Motorola)

curves for a MOSFET resemble the $V_{CE(sat)}$ curves of two-junction transistors. Figure 5–3 can be used as a design curve in much the same manner as the $V_{CE(sat)}$ curves of two-junction devices. For a given drain current (say, 2 mA), it is desired to know what value of V_{GS} is required to keep $V_{DS(ON)}$ below 0.5V. From the intersection of the I_D = 2 mA curve and the V_{DS} = 0.5V level, it is seen that V_{GS} must be greater than 5.3V to insure that V_{DS} will be below 0.5V.

Returning to Fig. 5–2, the voltage swing across the MOSFET capacitances can be determined. The voltage swing across C_{gs} is simply V_{IN}. The voltage change across $C_{d(sub)}$ is $V_{DD} - V_{DS(ON)}$. The voltage across C_{rss} is $V_{DD} + V_{IN} - V_{DS(ON)}$.

The variations in these capacitances, and r_{ds}, are responsible for the MOSFET switching behavior. The r_{ds} versus gate-source voltage and temperature for a typical MOSFET is shown in Fig. 5–4. Channel resistance decreases as V_{GS} is increased. This figure shows the variation of r_{ds} with temperature (−55°C to +125°C) to be about 0.15 ohm per degree C at a V_{GS} of 10V. This is comparable to about 3 ohms per degree C for a JFET.

C_{iss} and c_{rss} for a MOS device inverter are essentially independent of bias voltages, as shown in Fig. 5–5. The dip in the c_{iss} curves occurs at the start of conduction in the channel. $C_{d(sub)}$ (not shown) is a junction capacitance, and does vary (roughly) with the square-root of V_{GS}.

The switching curves for a typical MOSFET are given in Fig. 5–6. For this device, the gate time constant ($R_T C_{gs}$) is short compared to the drain time constant. Therefore, during the turn-on time, the channel is ON for most of the switching interval, resulting in shallow t_{d1} (turn-on delay time) versus I_D curves. That is, the turn-on delay time does not vary much over a wide range of I_D (7 to 43 nS for variations in I_D of 0.5 to 10V). The gate and drain time constants do not track over the current range. As a result, a decrease of the driving resistance R_S (as I_D increases) appears in the delay and rise time curves.

During turn-off, however, the channel is essentially OFF, and the charge time is determined by R_S and R_D alone. Thus, the curves of t_{d2} (turn-off delay time) and t_f (fall time) vary directly with the drain current.

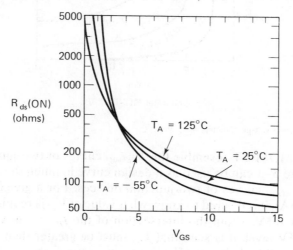

Fig. 5–4 2N4351 drain-source ON resistance. (Courtesy Motorola)

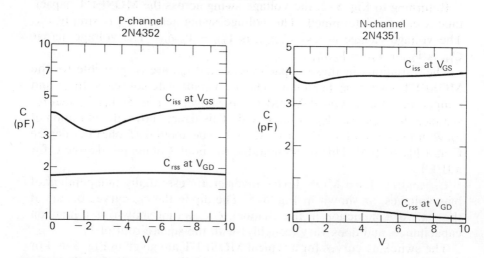

Fig. 5–5 C_{iss} and C_{rss} for MOS device inverter. (Courtesy Motorola)

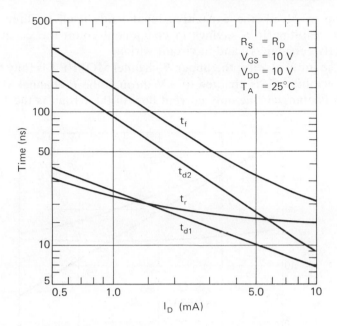

Fig. 5-6 2N4351 switching times. (Courtesy Motorola)

To emphasize the effect of the driving-source resistance R_S on switching times, the results of turn-on switching times (measured with very low source-resistance) are shown in Fig. 5-7. The turn-on delay time t_{d1} is most drastically affected, while the rise time at low I_D shows a similar improvement. At higher drain-current levels, the source resistor used for the $R_S = R_D$ condition is low, and improvement is not as great.

Summarizing the switching characteristics of MOSFETs are quite difficult to predict, at best, since they depend on the voltage variations of several characteristics, some of which are interrelated. A reasonable approach to giving design information on MOS logic devices is *to specify the switching characteristics under a variety of conditions.*

5-2. BASIC COMPLEMENTARY LOGIC

Although MOS logic is not limited to the complementary inverter, the complementary principle forms the backbone of most present-day MOS logic ICs, both standard and custom. The basic complementary inverter circuit is depicted in Fig. 5-8. This circuit has the unique advantage of dissipating almost no power in *either stable state.* Power is dissipated only during the switching interval. And, since MOS-

FETs are involved, the capacitive input lends itself to direct-coupled circuitry, resulting in a savings in component count (no capacitors required between stages), and in circuit wiring.

With the input at zero, the upper *P*-channel MOSFET is fully ON. The load capacitance C_L is charged to $+V$ through the *P*-channel MOSFET. Once C_L is charged, the only current flow in the circuit is the I_{DSS} of the

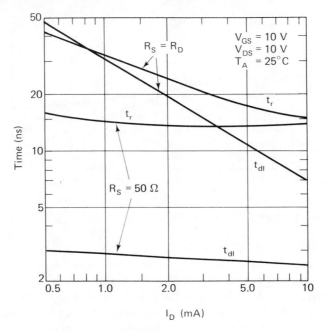

Fig. 5–7 2N4351 turn-on time variations. (Courtesy Motorola)

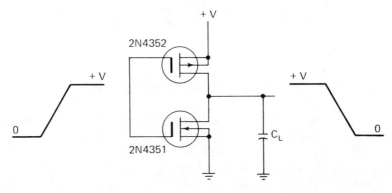

Fig. 5–8 Basic MOS device complementary inverter circuit. (Courtesy Motorola)

P-channel. Typically, this is in the picoampere range, since the MOSFET is completely OFF with zero gate voltage.

The voltage drop across the *P*-channel MOSFET is simply the I_{DSS} of the *N*-channel MOSFET multiplied by the channel resistance of the *P*-channel. Thus, if the I_{DSS} is 1 pA, and the channel resistance is 300 ohms, the voltage drop is about 3 nV.

With the input at +V, the *N*-channel is fully ON, and the *P*-channel is OFF. C_L discharges to ground and the *P*-channel MOSFET limits current flow to a few picoamperes. The voltage drop across the *N*-channel MOS-FET similarly is in the nanovolt region.

Figure 5–9 shows the switching performance of the complementary inverter of Fig. 5–8 as a function of fan-out. Here, fan-out is defined as C_L/C_{in}, where C_{in} for the complementary pair is about 10 pF. Thus, the rise time with a fan-out of 10 at $T_A = 25°C$ is 70 nA, and the delay time is 13 nS. At a fan-out of 100, representing a load capacitance C_L of 1000 pF, the rise time is about 650 nA, with a 100 nS delay time.

The temperature variation of the complementary inverter is also shown in Fig. 5–9. At a fan-out of 10, the variation in rise time over the temperature range of −40°C to +125°C is only 20 nA. There is no appreciable change in the delay time over this temperature range. Turn-off is much the same as turn-on, since one device of the pair is always turning on.

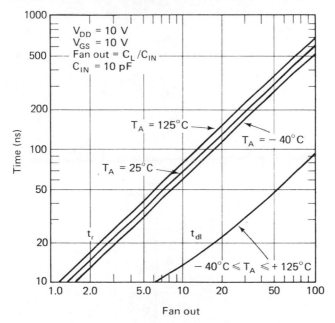

Fig. 5–9 Complementary inverter turn-on time temperature variation. (Courtesy Motorola)

5-2.1 Complementary NAND Gate

The complementary NAND gate is formed as shown in Fig. 5–10. The P-channel devices are connected in parallel, and the N-channel complements are connected in series. The truth table for the 3-input NAND gate is also given on Fig. 5–10. For the NAND function, the output is always high unless all three inputs are high. If any one or any pair of inputs are high, one or more of the P-channel devices will be held ON by the remaining low inputs, and the common output bus will be at +V. When all three inputs are high, all three series N-channels will be ON, and the output is low.

Note that the zero output level is developed across three series elements. However, the leakage current from all three of the P-channel devices is in the pA range, resulting in nV output levels (even for very large gates). For example, assume a P-channel leakage of 20 pA, and an N-channel resistance of 130 ohms. For a 50-input gate, the total leakage current is 1 nA (50 × 20 pA). Assuming a series output resistance of 6.5K

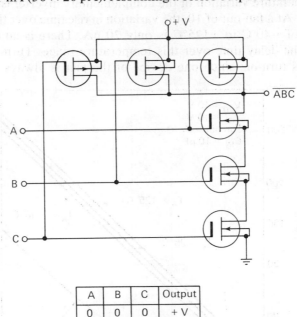

A	B	C	Output
0	0	0	+ V
0	0	1	+ V
0	1	0	+ V
0	1	1	+ V
1	0	0	+ V
1	0	1	+ V
1	1	0	+ V
1	1	1	0

Fig. 5–10 Three input NAND gate using MOS devices. (Courtesy Motorola)

(a high output resistance), resultant output voltage is $6.5\mu V$ (an extremely low output voltage).

As with any solid-state logic device, the limitation on width of the NAND gate is, of course, caused by decreasing switching speeds and increasing power dissipation as the width increases.

5-2.2 Complementary NOR Gate

The complementary NOR gate is shown in Fig. 5-11. Here, the order has simply been reversed. The P-channel devices are connected in series and the N-channels are in parallel. If any one of the inputs is high, one of the parallel N-channels will be ON, and the output will be low. Only when both inputs are low will both series P-channels be ON, allowing the output to become high. Thus, the conditions stated in the truth table are satisfied. The same comments regarding size of the NAND gate apply to the NOR gate.

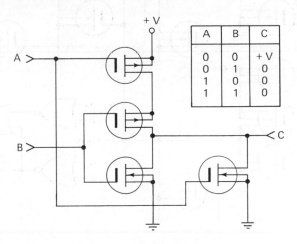

A	B	C
0	0	+ V
0	1	0
1	0	0
1	1	0

Fig. 5-11 Complementary NOR gate using MOS devices. (Courtesy Motorola)

5-2.3 Complementary Half-Adder

A practical example of how the complementary NOR gate can be used, Fig. 5-12 shows a half-adder using three NOR gates. The carry digit is taken from the NOR gate handling the complemented inputs, while the sum digit is picked up from the NOR gate that sums outputs from the first two NOR gates.

Referring back to Fig. 5-9, which shows the switching times of the complementary pair, consider the carry digit to the next stage as a fan-out of 1, and the input to the sum digit NOR gate as a fan-out of 1. The carry

NOR gate now faces a fan-out of 2. For a fan-out of 2 (according to Fig. 5–9), the rise time at room temperature should be 15 nA. Thus, the propagation delay from the output to the carry digit is 15 nS.

Assume a fan-out of 2 for the sum output NOR gate, and a total propagation delay from the input to the sum digit of 30 nS. This allows 15 nA to the carry input, plus another 15 nA through the final NOR gate, for a total of 30 nS propagation delay.

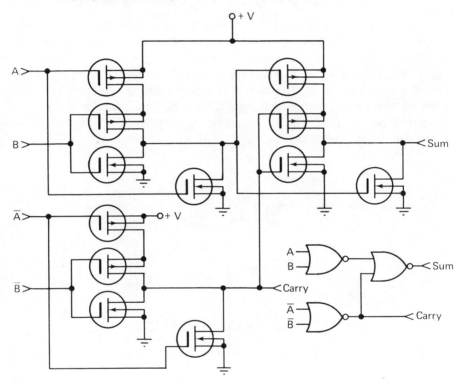

Fig. 5–12 Half adder using complementary MOS device NOR gates. (Courtesy Motorola)

5–2.4 Modified Half-Adder

The half-adder can be modified to handle only one set of inputs by changing one NOR gate to an AND gate, as illustrated by Fig. 5–13. (The AND gate is formed by a NAND gate followed by an inverter, as described in the author's *Handbook of Logic Circuits*.)

The additional stage increases the propagation delay somewhat. The fan-out from the NAND gate is 1 for a propagation delay of 10 nS, and the fan-out of the inverter is assumed to be 2 for an additional 15 nS. Total propagation time to the carry digit is now 25 nS, and to the sum digit, 40

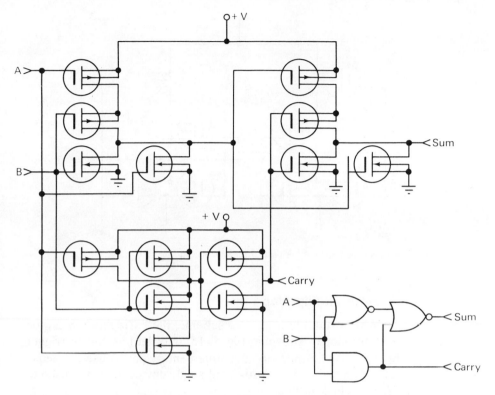

Fig. 5-13 Half adder (not requiring complementary inputs) using MOS devices. (Courtesy Motorola)

nS. The additional stage adds 10 nS to the propagation delay times expected in the circuit of Fig. 5-12.

5-2.5 *R-S* Flip-Flop

An *R-S* flip-flop implemented from complementary MOS devices is somewhat more complex than simply cross-coupling gates. One version of the *R-S* flip-flop is given in Fig. 5-14.

One of the major advantages of MOSFET logic is the ability to design highly reliable direct-coupled circuits. However, care must be exercised to avoid *sneak-paths* that are an inadvertent hazard of direct-coupled circuits. The *R-S* flip-flop is a case in point.

If the set and reset lines apply directly to the basic flip-flop gates, the output is driven directly from the set and reset lines. An additional complementary pair is required for both the set and reset lines to isolate them from the load. The circuit of Fig. 5-14 then becomes the basic flip-flop circuit for complementary MOS logic.

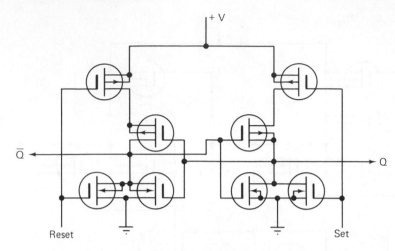

Fig. 5-14 Complementary MOS device *R-S* flip-flop. (Courtesy Motorola)

5-2.6 *JK* Flip-Flop

With a suitable triggering scheme, the basic flip-flop can be made into a *JK* flip-flop as shown in Fig. 5–15. In order to steer the trigger pulse for the $J = K = 1$ condition, a complementary pair is used to sense the output of the opposite side. Additional single device gates are used to gate the *J*, *K* and clock inputs.

Assume that the *J* output is grounded, and that the $J = K = 1$ condition is to toggle the flip-flop. With the *J* output low, the *P*-channel of the right-hand sensing pair is biased ON. High inputs at the *J* and *K* inputs turn on their respective gates. When the clock input is high (true), the $+V_{DD}$ is propagated through the clock gate, the *K* input, the right-hand sensing gate, and to the *K* output gate. This toggles the flip-flop.

The trigger signal does not propagate through the left-hand sensing gate since it is blocked. This scheme requires a *narrow clock pulse*. The clock pulse must not be present when the *K* output has switched since the left-hand sensing pair would then be biased ON.

The circuit of Fig. 5–15 can be operated at frequencies up to about 5 MHz without undue precautions. A frequency of about 10 MHz is possible with careful layout and component selection.

5-3. STATE-OF-THE-ART DIGITAL IC MOS DEVICES

In this section we shall describe three present-day lines of standard digital MOS devices. These include: the McMOS, which is the trademark of Motorola Semiconductor Products, Inc.; complementary

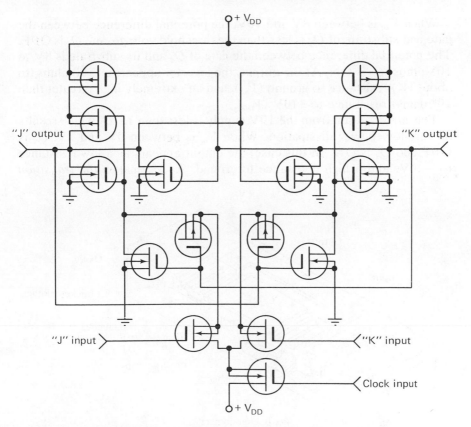

Fig. 5-15 Complementary *J-K* flip-flop using MOS devices. (Courtesy Motorola)

MOS devices; the COS/MOS, which is the Radio Corporation of America complementary MOS line; and the standard MOS/LSI line of Texas Instruments Inc.

Keep in mind that these descriptions are for present-day equipment and are subject to change. That is, the details such as power requirements, propagation times, logic levels, etc. may change in the future. However, the *basic principles for MOS devices,* both standard lines and custom products, will remain the same. A careful study of this section will enable the reader to interpret the future data of the manufacturers represented, as well as the data of other manufacturers.

5-3.1 Motorola Complementary MOS ICs

Figure 5-16 shows how the *P*-channel and *N*-channel devices are connected to form the basic element of McMOS. Figure 5-17 gives the V_{in} versus V_{out} transfer curve for the basic inverter at a power supply voltage of 10V.

When V_{in} is between 8V and 10V, the potential difference between the gate and substrate of Q_1 is less than the threshold voltage, and Q_1 is OFF. The potential difference between the gate of Q_2 and its substrate is 8V to 10V; thus, Q_2 is ON. At the output, the inverter appears as 500 ohms (to about 1K) resistance to ground (V_{SS}), and an extremely high (greater than 10^9 ohms) resistance to +10V (V_{DD}).

The current drain from the 10V supply is less than 15 nA which results in very-low power dissipation. When V_{in} is between 0V and 2V, Q_2 is OFF and Q_1 is ON. In this case, the output appears as a low resistance to +10V, and a high resistance to ground. As V_{in} makes the *transition*

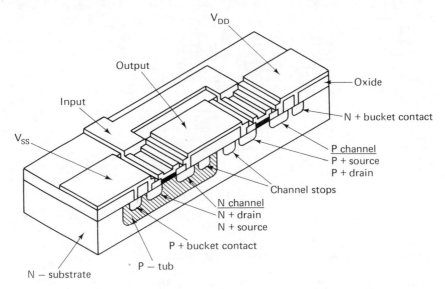

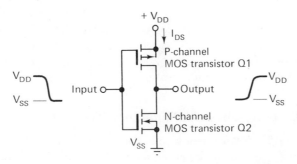

Fig. 5–16 Basic CMOS *P*-channel and *N*-channel devices connected to form an inverter. (Courtesy Motorola)

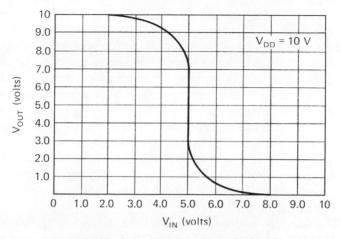

Fig. 5–17 McMOS inverter transfer curve. (Courtesy Motorola)

from 2V to 8V, both Q_1 and Q_2 are in an ON condition, resulting in current flow from V_{DD} to ground. Figure 5–18 shows the current as a function
of V_{in}.

When a switching waveform is applied to V_{in}, current I_D does not flow
through both Q_1 and Q_2. Assume Q_1 is OFF and Q_2 is ON. As V_{in}
switches from V_{DD} to ground, there is a period of time when both Q_1 and
Q_2 are ON. However, because of stray and load capacitance on the output, the output voltage is still nearly 0V when Q_1 and Q_2 are both ON.
Thus, little current flows through Q_2 to ground. Power dissipation of
McMOS circuits, under switching conditions, is caused almost entirely
by capacitive loading.

dynamic characteristics — Because dynamic power dissipation results

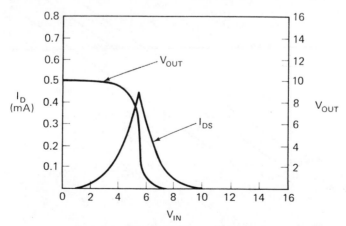

Fig. 5–18 Power supply current versus input voltage of McMOS inverter. (Courtesy Motorola)

from capacitive loading, dissipation is also a function of the frequency at which the capacitance is charged and discharged. Figure 5–19 illustrates this relationship for a basic inverter. It can be seen that power dissipation is linear with frequency. Figure 5–20 shows the effect of capacitive loading and power supply voltage on propagation delays. It is evident that higher operating speeds are possible at higher supply voltages at the expense of power dissipation.

The input capacitance is dependent on the input voltage, as depicted in Fig. 5–21. The input capacitance increases when the output is in the transition region. This increase is due to internal feedback. Under normal operating conditions, this effect is not present because the input transition is completed before the feedback takes place because of the propagation delay of the gate.

NAND and NOR functions—Figure 5–22 shows how the basic inverter is connected to form the NAND and NOR logic functions.

For the NOR function, the gates of Q_1 and Q_3 are tied together to form input 1 and a basic inverter. The gates of Q_2 and Q_4 form input 2 and a second inverter. Device Q_2 acts as a series resistance which is either extremely high or low, depending on its gate signal in the inverter formed by Q_1 and Q_3. Likewise, Q_1 acts as a series resistance in the second inverter. The output of the circuit is at +10V only when both Q_1 and Q_2 are ON, occurring only if both inputs 1 and 2 are at ground. Thus, the output is a logic "1" only when both inputs are logic "0" (which is the conventional NOR function).

The McMOS NOR gate can be converted to a NAND gate by inter-

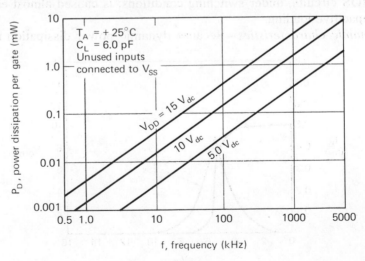

Fig. 5–19 Power dissipation of McMOS inverter. (Courtesy Motorola)

changing the *P*- and *N*-channel devices, and by turning the circuit upside down, as shown in Fig. 5–22. Both Q_3 and Q_4 must be ON for the output to be in the logic 0 condition.

More inputs can be added to the NAND or NOR functions to make 3- and 4-input gates by adding complementary pairs. Because of the series resistance and internal biasing, a practical limit is about 4 inputs.

gate transfer characteristics—Figure 5–23 shows the typical transfer characteristic curves of a McMOS gate. Two characteristics are especially evident. First, there is little change in transfer curves over the temperature range. Second, as the supply voltage changes, the threshold voltage changes. Typically, the threshold is about 45 percent of the supply voltage, resulting in the capability of complementary MOS logic to operate over a wide supply range. Since the output is equally isolated from both V_{DD} and V_{SS} terminals, McMOS can operate with negative as well

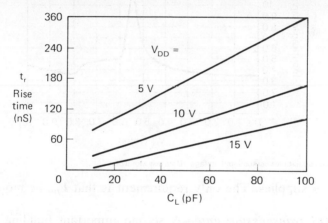

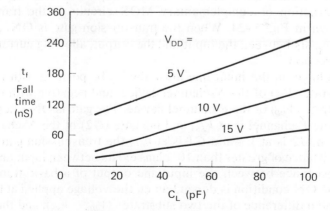

Fig. 5–20 Typical delay characteristics of 2-input McMOS NOR gate. (Courtesy Motorola)

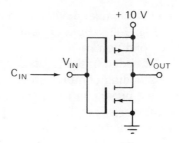

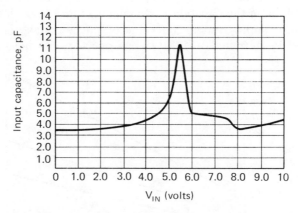

Fig. 5–21 Input capacitance versus input voltage. (Courtesy Motorola)

as positive supplies. The only requirement is that V_{DD} be more positive than V_{SS}.

McMOS transmission gate—A second important building block for the construction of complementary MOS circuits is the transmission gate shown in Fig. 5–24. When the transmission gate is ON, a low resistance exists between the input and the output, allowing current flow in either direction.

The voltage on the input line must always be positive with respect to the substrate (V_{SS}) of the N-channel device, and negative with respect to the substrate (V_{DD}) of the P-channel device. The gate is ON when the gate ($G1$) of the P-channel is at V_{SS}, and the gate ($G2$) of the N-channel is at V_{DD}. When $G2$ is at V_{SS} and $G1$ is at V_{DD}, the transmission gate is OFF, and a resistance of greater than 10^9 ohms exists between input and output.

The resistance between the input and output of a basic transmission gate in the ON condition is dependent on the voltage applied at the input, the potential difference of the two substrates ($V_{DD} - V_{SS}$), and the load on the output. R_{ON} is defined as the input-to-output resistance with a 10K

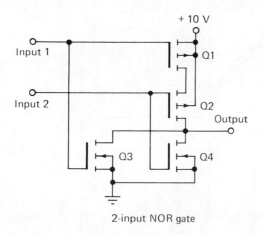

+ 10 V

"0" = 0 V		"1" = + 10 V
Input 1	Input 2	Output
0	0	1
0	1	0
1	0	0
1	1	0

2-input NOR gate

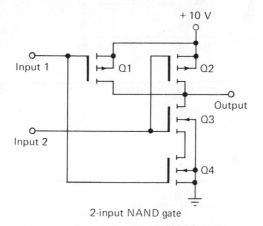

+ 10 V

"0" = 0 V		"1" = + 10 V
Input 1	Input 2	Output
0	0	1
0	1	1
1	0	1
1	1	0

2-input NAND gate

Fig. 5–22 Two-input McMOS gates. (Courtesy Motorola)

load resistor from the output to ground. Figure 5–25 illustrates an interesting peaking effect which occurs in the R_{ON} versus V_{in} curves of the basic transmission gate. When V_{in} is at, or near, V_{DD}, the P-channel device is providing the low resistance. The N-channel device is OFF since the potential difference between $G2$ and the drain or source of the N-channel device is less than the threshold voltage. When V_{in} is at or near V_{SS}, the N-channel is conducting and the P-channel is OFF.

With voltages between the two extremes, both devices are partially ON and the value of R_{ON} is caused by the parallel resistance of the P- and N-channel devices. The different slope of the curve on either side of the peak is due to the greater sensitivity of the N-channel resistance to substrate degeneration (or substrate bias) than the P-channel. Thus, the rate of

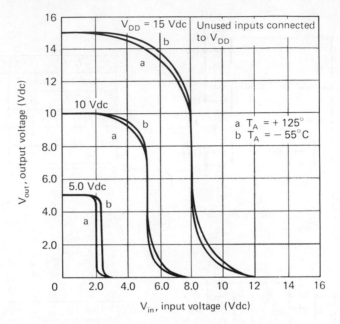

Fig. 5–23 Typical voltage transfer characteristics versus temperature for McMOS gate. (Courtesy Motorola)

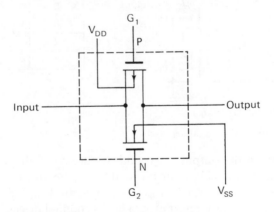

Fig. 5–24 Basic McMOS transmission gate. (Courtesy Motorola)

increase in R_{ON} with respect to V_{in} is greater for input voltages between V_{SS} and the *peaking voltage* than for input voltages greater than the peaking voltage.

modified transmission gate—Figure 5–26 shows a modification of the basic transmission gate. The modification is implemented by the addition of a third device to control the substrate bias of the N-channel device. The effect of the third device is to delay the turn-off of the N-channel device, resulting in a much flatter R_{ON} versus V_{in} curve as shown in Fig. 5–25.

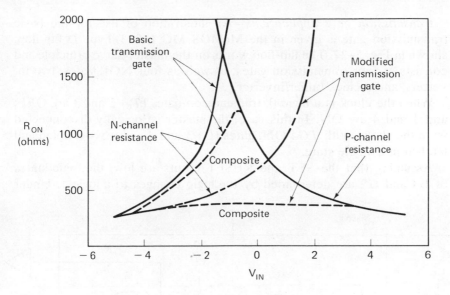

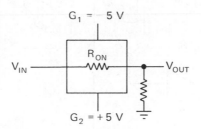

Fig. 5–25 McMOS transmission gates R_{ON} versus V_{in}. (Courtesy Motorola)

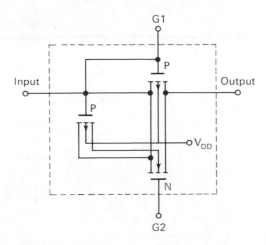

Fig. 5–26 Modified McMOS transmission gate. (Courtesy Motorola)

transmission gate applications—An illustration of use for the basic transmission gate is given in the McMOS MC 14013 Type *D* flip-flop, shown in Fig. 5–27. The flip-flop works on the master-slave principle and consists of four transmission gates, as well as four NOR gates, two inverters, and a clock buffer/inverter.

When the clock is a logic 0, transmission gates (T_G) 2 and 3 are OFF, and 1 and 4 are ON. In this case, the master is logically disconnected from the slave. With $TG4$ ON, gates (G) 3 and 4 are cross-coupled and latched in a stable state.

Assuming that the SET and RESET inputs are low, the logic states of $G1$ and $G2$ are determined by the logic changes to a logic 1. Under

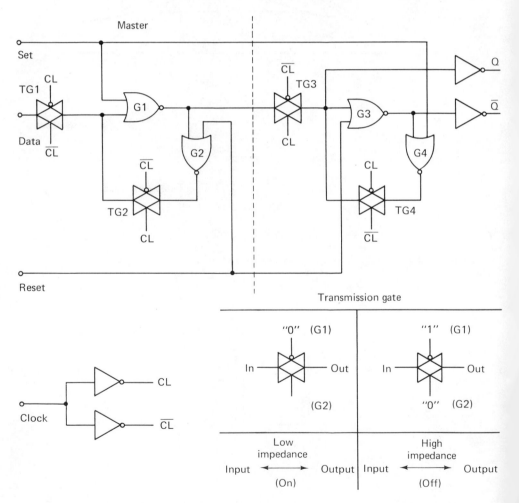

Fig. 5–27 McMOS type-*D* flip-flop. (Courtesy Motorola)

these conditions, $TG2$ and $TG3$ turn ON, and $TG1$ and $TG4$ turn OFF. Gates $G1$ and $G2$ are cross-coupled through $TG2$, and the gates latch into the state in which they existed at the time the clock changed from a 0 to a 1. With $TG3$ ON, the logic state of the master section (output of gate $G1$) is fed through an inverter to the Q output, and through $G3$ and another inverter to the $\overline{Q}$ output.

When the clock returns to a logic 0, $TG3$ turns OFF, and $TG4$ turns ON. This disconnects the slave from the master, and latches the slave into the state that existed in the master when the clock changed from a 1 to a 0. Thus, data is entered into the master on the positive edge of the clock. When the clock is high, the output of the master is transmitted directly through the slave to Q and $\overline{Q}$. When the clock changes back to a low state, the state of the master is stored by the slave which then provides the output.

input protection circuits—The input protection circuits for the McMOS devices are described in Sec. 1–5.5 of Chap. 1.

5–3.2 RCA COS/MOS Complementary MOS Devices

COS/MOS fundamentals can best be understood by reference to Fig. 5–28. As shown, the MOS enhancement mechanism is a majority-carrier device. The current in a conducting channel between two diffused electrodes (source and drain) is controlled (enhanced) by a voltage applied to a third terminal (gate). The gate is insulated from the source and drain.

In an N-type device, the majority carriers are electrons. A positive voltage on the gate is required to enhance the conducting channel. For all gate voltages less than a threshold value (V_{th}), the conductivity of the

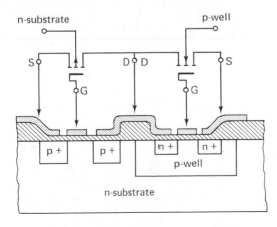

Fig. 5–28 Cross-section of COS/MOS device. (Courtesy RCA)

channel is negligible and the device is said to be cut off. For gate voltages greater than V_{th}, the channel is enhanced, and current flow in the channel occurs if a suitable voltage is applied between the source and drain. The resultant device characteristics are shown in Fig. 5–29(a).

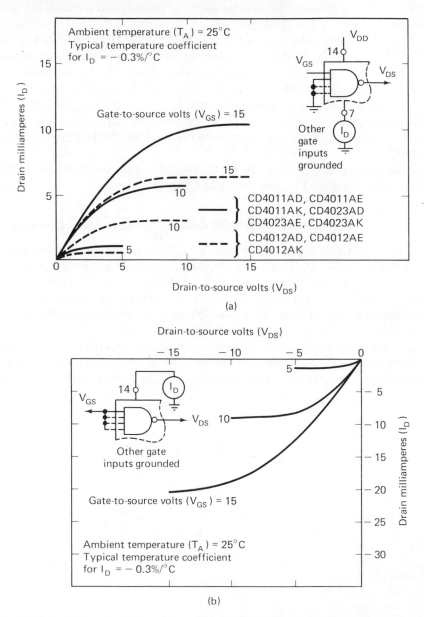

(a)

(b)

Fig. 5–29 Typical *N*-channel and *P*-channel characteristics of COS/MOS devices. (Courtesy RCA)

Operation of the *P*-type device is essentially the same as that for the *N*-type, except that the carriers are holes and the applied voltage required to enhance the channel must be negative rather than positive. [See Fig. 5–29(b).]

The gate electrode for a device of either polarity is insulated from the body of the device. As a result, current flows only from source to drain in the channel, never from the gate into the channel.

quiescent Device Dissipation—The basic logic inverter (or logic gate) formed by use of a *P*- and *N*-type device in series is illustrated in Fig. 5–30. When the input lead is grounded, or otherwise connected to 0V (logical 0), the *N*-device is cutoff and the *P*-device is biased on. As a result, there is a low-impedance path from the output to V_{DD}, and an open circuit to ground. The resultant output voltage is essentially V_{DD}, or a logic 1.

When this occurs, the *N*-channel device becomes a low impedance, while the *P*-channel device becomes an open circuit. The resultant output becomes essentially zero volts (logic 0).

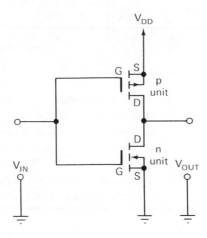

Fig. 5–30 Basic COS/MOS inverter. (Courtesy RCA)

Note that one of the devices is always cut off at either logic extreme, and that no current flows into the insulating gates, resulting in negligible inverter quiescent power dissipation (equal to the product of V_{DD} times the leakage current).

A cross section of the COS/MOS inverter (as it is formed in an integrated circuit) on an *N*-type substrate is illustrated in Fig. 5–28. Compare this with Fig. 5–31. Note that the source-drain diffusions and the *p*-well diffusion form parasitic diodes (in addition to the desired transistors) at the basic inverter nodes. (Keep in mind that the parasitic diodes are not to be confused with the protective diodes, described in later paragraphs.)

The parasitic diode elements are back-biased (across the power supply) and contribute, in part, to the device leakage current, and thus to the quiescent power dissipation.

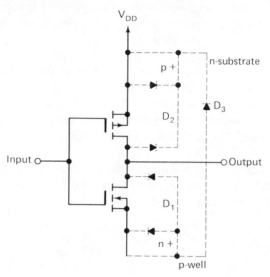

Fig. 5–31 Basic inverter showing parasitic diodes. (Courtesy RCA)

The RCA COS/MOS product line consists of circuits of varying complexity (from the dual 4-input logic gates that contain 16 MOS devices, to the more complex 64-bit static shift registers that contain over 1000 devices). The COS/MOS devices occupy different amounts of silicon area, and are composed of varying numbers of circuits formed from inverters. Consequently, each device in the family shows a particular magnitude of leakage current, depending on the total effect of device count and parasitic diode area.

For example, some logic gates are specified to operate with a typical power dissipation of 5 nW ($V_{DD} = 10$V), but 7-stage counters or registers are specified to operate with a typical power dissipation of 5 μW ($V_{DD} = 10$V). Published data includes both typical device quiescent-current levels and maximum levels ($V_{DD} = 5$V and $V_{DD} = 10$V).

switching characteristics—The input/output characteristics for the COS/MOS inverter area are shown in Fig. 5–32. The signal extremes at the input and output are approximately zero volts (logic 0) and V_{DD} (logic 1). The switching point is shown to be typically 45 to 55 percent of the magnitude of the power supply voltage (regardless of the magnitude) over the entire range from 3 to 15V (or 5 to 15V). Note the negligible change in operating point from −55°C to +125°C.

ac dissipation characteristics—During the transition from a logic 0 to

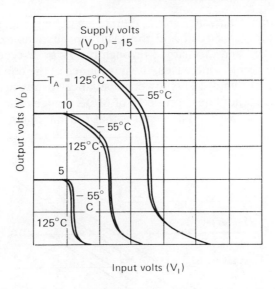

Fig. 5–32 Typical COS/MOS transfer characteristics as a function of temperature. (Courtesy RCA)

a logic 1, both devices are momentarily ON. This condition results in a pulse of instantaneous current being drawn from the power supply, the magnitude and duration of which depends on the following factors:

1. The impedance of the particular devices being used in the inverter circuit.
2. The magnitude of the power-supply voltage.
3. The magnitude of the individual device threshold voltages.
4. The input driver rise and fall times.

An additional component of current must also be drawn from the power supply to charge and discharge the internal parasitic node capacitances and the load capacitances seen at the output.

The device power dissipation resulting from these current components is a frequency-dependent parameter. The more often the circuit switches, the greater the resultant power dissipation; the heavier the capacitive loading, the greater the resultant power dissipation. The power dissipation is not duty-cycle dependent. For practical purposes, power dissipation can be considered frequency (repetition-rate) dependent.

Because the COS/MOS product line ranges widely in circuit complexity from device to device, the ac device dissipations vary widely. The effect of capacitive loading on the individual devices also varies. Figure 5–33 shows a family of curves for a typical gate device, and a typical MSI (medium-scale integration) device. These curves illustrate how device power dissipation varies as a function of frequency, supply voltage, and capacitive loading.

ac performance characteristics—During switching the node capacitances within a given circuit and the load capacitances external to the circuit are charged and discharged through the *P*- or *N*-type device conducting channel. As V_{DD} increases, the impedance of the conducting channel decreases accordingly. This lower impedance results in a shorter RC time constant (this nonlinear property of MOS devices can be observed in the curves of Fig. 5–29). The result is that the maximum switching fre-

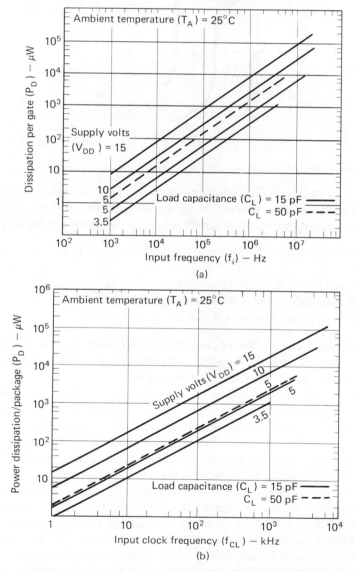

(a)

(b)

Fig. 5–33 Typical power dissipation characteristics: (a) basic gate power dissipation characteristics; (b) MSI device power dissipation characteristics. (Courtesy RCA)

quency of a COS/MOS device increases with increasing supply voltage (see Fig. 5–34).

Figure 5–34(b) shows curves of propagation delay as a function of supply voltage for a typical gate device. However, the trade-off for low supply voltage (lower output current to drive a load) is lower speed of operation.

calculating system power—The following guidelines have been developed to assist the designer in estimating system power for the COS/MOS

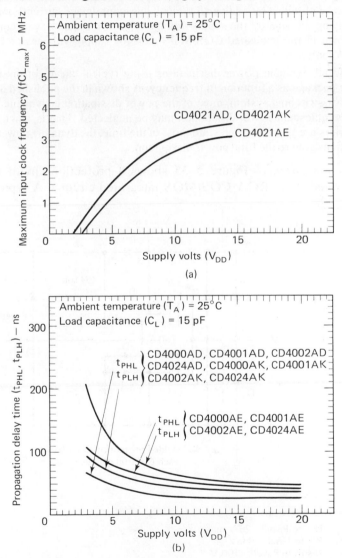

(a)

(b)

Fig. 5–34 Operating frequency and propagation delay as a function of power-supply voltage; (a) maximum guaranteed operating frequency as a function of power-supply voltage; (b) propagation delay as a function of power-supply voltage for the basic gate. (Courtesy RCA)

line. The same general guidelines can be applied to similar MOS lines.

Total system power is equal to the sum of quiescent power and dynamic power. Therefore, system power can be calculated with the following two-step approach:

1. Add all typical package quiescent power dissipations, using published data. Because quiescent power dissipation is equal to the product of quiescent device current multiplied by supply voltage, quiescent power may also be obtained by adding all typical quiescent device currents, and multiplying the sum by the supply voltage (V_{DD}). Quiescent device current is shown in the published COS/MOS data for supply voltages of 5V and 10V only.

2. Add all dynamic power dissipations using typical curves of dissipation per package as a function of frequency, as shown in the published data. In a fast-switching system, most of the power dissipation is dynamic, therefore, quiescent power dissipation may be neglected. That is, since the inverters are in a transition state most of the time, the dynamic power dissipations govern the total power dissipation.

protection circuits—Figure 5–35 shows a protection circuit which is incorporated in all RCA COS/MOS integrated circuits. A typical value

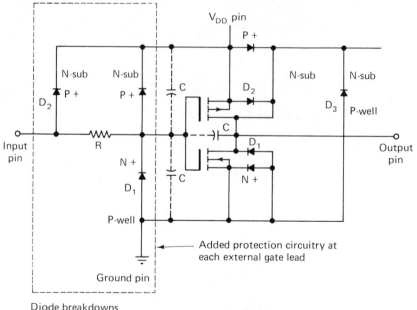

Diode breakdowns

D_1 = N + to P-well   25 V max
D_2 = P + to N-sub   50 V
D_3 = N-sub to P-well   100 V
R   = Normal P + diffusion in N-sub isolation

Fig. 5–35 Gate-oxide protection circuit used in COS/MOS integrated circuits.(Courtesy RCA)

of 500 ohms is used for the input resistor R because this value, in conjunction with the capacitance of the gate and the associated protective diodes, integrates and clamps the device voltages to a safe level.

The diagrams of Fig. 5–36 show how all circuits are protected under various operating conditions. These conditions include protection between input and V_{DD}, between input and ground, and between input and output.

5–3.3 Texas Instruments MOS/LSI Products

The following descriptions apply to the Texas Instruments MOS/LSI line. This line includes shift registers, read-only memories (ROM), programmable logic arrays (PLA), and random access memories (RAM), as well as special purpose devices such as buffers, switches and custom MOS/LSI devices.

5–3.3.1 *Shift Registers*

For all digital equipment, there is a need to temporarily store and transfer data. MOS shift registers are ideally suited for these applications, because MOS devices economically can store very large amounts of information.

basic configuration—MOS shift registers can be supplied in the following configurations: Serial-in/serial-out, parallel-in/serial-out, and serial-in/parallel-out. The *serial-in/serial-out* configuration is by far the most popular.

A MOS shift register will be able to store N bits, each on a *basic cell* consisting of two MOS inverters and timing devices, as shown in Fig. 5–37.

static versus dynamic—Dynamic shift registers use two independent inverters (not cross-coupled). The data is stored temporarily on a capacitor inherent to the MOS device (gate capacitance). The device cannot be operated below a certain clock frequency or the data storage will be lost.

A *static shift register* (Fig. 5–38) operates in the same way as a dynamic shift register, as long as the frequency is high. The two inverters used in a static shift register are the static type (unclocked load). When the frequency falls below a certain level, a third phase is generated internally, and this signal is used to close a feedback loop between the output of the second inverter and the input of the first inverter.

Comparing the two, we see that dynamic shift registers are faster and use less power than static shift registers. However, dynamic shift registers are not as flexible to use in a system.

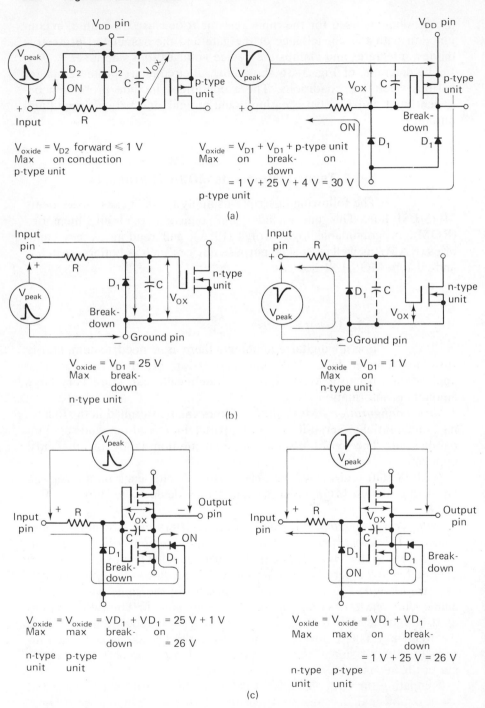

V_{oxide} = V_{D2} forward ≤ 1 V
Max   on conduction
p-type unit

V_{oxide} = V_{D1} + V_{D1} + p-type unit
Max   on   break-   on
                    down
        = 1 V + 25 V + 4 V = 30 V
p-type unit

(a)

V_{oxide} = V_{D1} = 25 V
Max   break-
              down
n-type unit

V_{oxide} = V_{D1} = 1 V
Max   on
n-type unit

(b)

V_{oxide} = V_{oxide} = VD_1 + VD_1 = 25 V + 1 V
Max   max   break-   on
                          down   = 26 V
n-type   p-type
unit   unit

V_{oxide} = V_{oxide} = VD_1 + VD_1
Max   max   on   break-
                                    down
                          = 1 V + 25 V = 26 V
n-type   p-type
unit   unit

(c)

Fig. 5–36 Circuits used to provide; (a) protection between input pin and V_{DD} pin; (b) protection between input pin and ground pin; (c) protection between input pin and output pin. (Courtesy RCA)

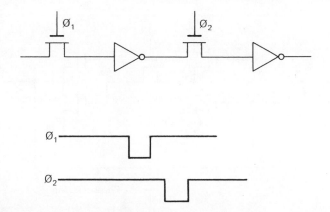

Fig. 5-37 Basic MOS/LSI shift register cell. (Courtesy Texas Instruments)

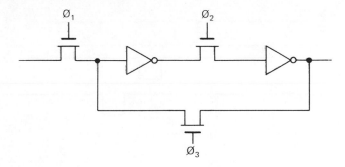

Fig. 5-38 Basic MOS/LSI static shift register. (Courtesy Texas Instruments)

static shift registers—As illustrated in Fig. 5–39, a static shift register uses two static MOS inverters. Three phases (clocks) are necessary for operation. The third-phase clock is *always generated internally,* and is used to time the feedback loop. The second-clock phase is often generated internally.

In the basic cell of Fig. 5–39, *A* and *B* are storage nodes. The device operates dynamically except when phase 3 is on. Phase 3 is present only when phase 1 is at logic 0 and phase 2 is logic 1 for more than 10 microseconds. This condition must be maintained for long-time data storage. As shown in Fig. 5–40, phase 3 is actually a delayed phase 2, generated by an inverter. The load devices associated with the static shift register remain on at all times.

Static shift registers typically operate in the 0 to 2.5 MHz clock range. Static registers are extremely flexible, and data can be held indefinitely (as long as power is supplied).

dynamic shift registers—Dynamic shift registers use either two or four phases (clocks). These phases can be generated on the chip or supplied externally. Two-phase shift registers can be classified as *ratio* and *ratioless* circuits.

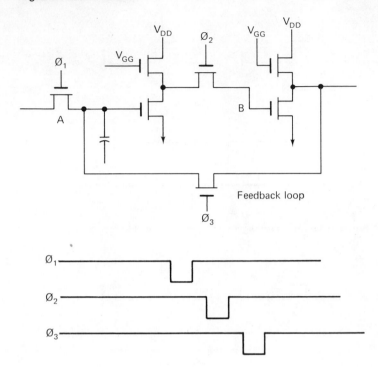

Fig. 5-39 Basic cell of MOS/LSI static shift register. (Courtesy Texas Instruments)

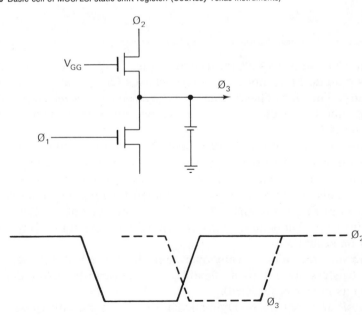

Fig. 5-40 Generation of $\phi3$ in MOS/LSI register. (Courtesy Texas Instruments)

The *two-phase ratio-type shift register* (Fig. 5–41) consists of two simple dynamic inverters and timing devices. When phase 1 is at a logic 1 (low), the capacitance C_1 charges at the inverse of the data input. Information is transferred out when phase 2 goes to 1.

In a ratio-type circuit, current flows through the inverter when the clock and data input are at a logic 1 simultaneously. There must be a certain minimum ratio between the size of the two transistors in the inverters (typically 5 to 1), requiring more chip area than in a ratioless shift register in which the MOS devices are usually identical in size.

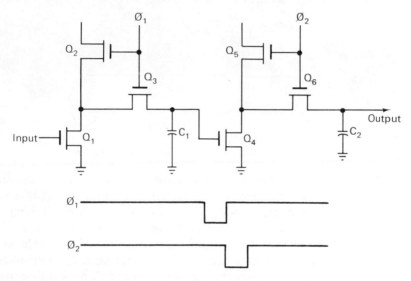

Fig. 5–41 Basic cell for MOS/LSI dynamic shift register. (Courtesy Texas Instruments)

The *two-phase ratioless dynamic shift register* (Fig. 5–42) has been designed to decrease the power dissipation and chip area. This ratioless register uses identical transistors throughout, and can therefore work at higher clock rates because the precharging paths are of lower impedance than those in the circuit.

When phase 1 goes to 1, C_2 charges to 1 via Q_3, and C_1 charges to the data input level via Q_1. When phase 1 returns to 0, Q_2 turns on if the input level was a 1. This discharges C_2.

For a 0 input, Q_2 stays off and C_2 is not discharged. Under these conditions, phase 2 goes to a 1 and turns on Q_4 so that C_2 shares any charge it has with C_4. Capacitor C_3 is used to compensate for the loss of potential across C_2 by introducing a small extra charge on the negative edge of phase 2. However, the small charge does not introduce enough energy to destroy a logic 0 on C_2. When phase 2 returns to a 0, the charge on C_4 transfers the data-input level to the output.

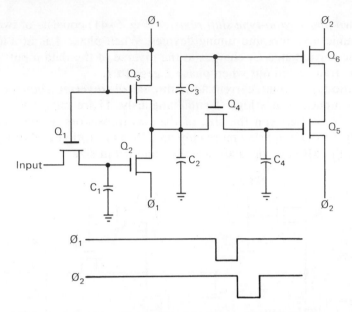

Fig. 5–42 MOS/LSI two-phase ratioless dynamic shift register. (Courtesy Texas Instruments)

Four-phase shift registers (Fig. 5–43) are used for very high density circuits operated at very high speed. In the basic four-phase dynamic shift register, C is precharged via Q_1 during phase 1. After phase 1, phase 2 holds Q_2 on, so C takes a level which is the inverse of the input. The process is repeated by the slave section $Q_4 - Q_6$ so that the input level is transferred to the output after phase 3 and during phase 4. The four-phase register uses similar transistors throughout, giving high package density. Power dissipation is low, speed can be high, but a relatively complex clock drive (four phases) is required.

clocking the shift registers—Most of the new shift registers designed by Texas Instruments feature total *TTL* (transistor-transistor logic) compatibility. This is accomplished by including on the chip a clock driver driven directly by TTL circuits. In older MOS shift registers, the user must generate the clocks, and shift their level (from TTL to MOS).

When the MOS shift register does not have direct TTL compatibility, interfacing can be accomplished with a *D*-type flip-flop, as shown in Fig. 5–44. This arrangement is favorable because Q and $\overline{Q}$ will be of opposite polarities when the TTL clock is stopped, allowing static storage in static registers. Two flip-flops can be used when four clock phases are required (such as with the register of Fig. 5–43). The flip-flops must be capable of shifting the level to that suitable for the MOS register. The subject of interfacing MOS logic with other logic forms is discussed in Sec. 5–4.

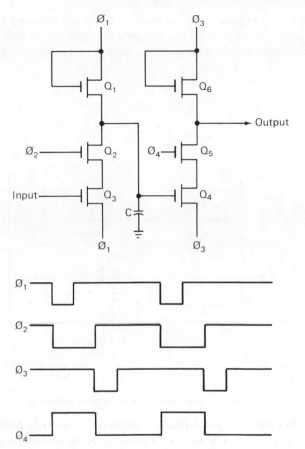

Fig. 5–43 MOS/LSI four-phase shift register. (Courtesy Texas Instruments)

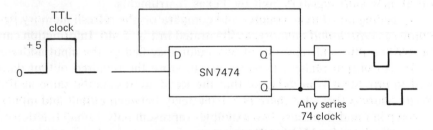

Fig. 5–44 Clocking MOS/LSI from TTL. (Courtesy Texas Instruments)

typical applications—Typical applications for MOS/LSI shift registers include: data handling, refresh memories, buffer memories, scratch-pad memories, delay line, desk-top calculators, display systems, peripherals, and radar systems. The following is a brief description of some typical memories using basic MOS shift registers.

Any *N*-bit shift register can be used as a *refresh memory* by returning outputs to inputs as in Fig. 5–45. The rate (in seconds) at which a particular bit of information is available at the output is determined by the equation:

$$\frac{N}{\text{clock frequency}}$$

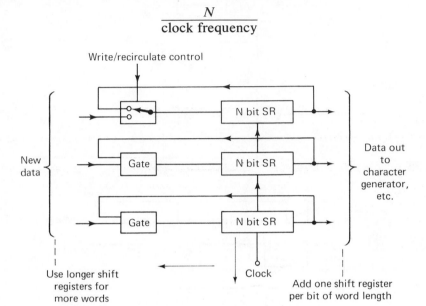

Fig. 5–45 MOS/LSI shift register used as refresh memory. (Courtesy Texas Instruments)

This arrangement is particularly useful for renewing fading displays such as CRT character-generator systems. New information is written in via a 2-way input gate circuit as shown. This gate circuit is incorporated on all new shift registers built by Texas Instruments.

By adding an address counter and comparator, the refresh memory becomes a *scratch-pad memory,* as illustrated in Fig. 5–46. Information can be written into and read out of any point specified by the input address code. An output register is necessary to store the required output data and to provide a 1-bit delay so that the Read address is the same as the Write address (because there is a 1-bit delay between output and input).

Keep in mind that these two examples represent only a small fraction of the applications for MOS shift registers.

5–3.3.2 *Read-Only Memories*

The information stored in a read-only memory (ROM) is permanently programmed into the circuit at the time of its manufacture. Once the information is entered, it cannot be changed. However, the in-

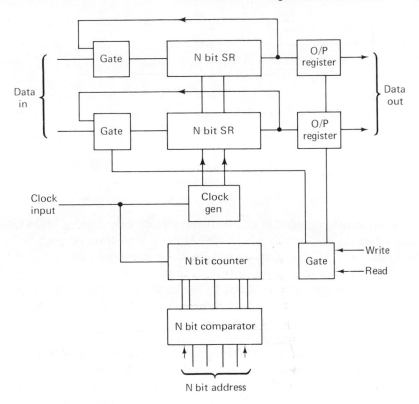

Fig. 5–46 MOS/LSI shift registers used as scratch pad memory. (Courtesy Texas Instruments)

formation can be read out as often as desired. Before MOS logic circuits became available, the only practical means of realizing a ROM were with discrete-diode matrices, or core memories. The most obvious advantages of MOS ROMs are:

cost MOS typically one-tenth that of a diode matrix

size MOS can put 4096 bits in a 24-pin package (chip size is 120×110 mil)

speed New MOS techniques can provide access times as low as 50 nS.

A single MOS ROM device will be made up of three sections (Fig. 5–47):

decoder in which the binary address is decoded and $X-Y$ pairs of lines going to the memory matrix are enabled (one pair of $X-Y$ lines if there is one bit per output word; two pairs of $X-Y$ lines if there are two bits per output word, etc.)

memory matrix containing as many MOS *transistor locations* as there are bits in the memory.

buffer which supplies output levels for the external circuitry.

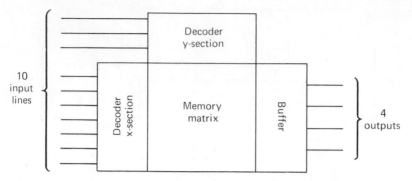

Fig. 5–47 Single MOS/LSI read only memory (ROM) sections. (Courtesy Texas Instruments)

The physical arrangement of the memory matrix containing MOS transistor locations is shown in Fig. 5–48. At the intersection of every X-line and Y-line, any MOS transistor can be either constructed or omitted by growing either a thin-gate oxide or a thick-gate oxide. The absence of an MOS transistor will be interpreted by the buffer as a logic 0, and the presence of a thin-gate MOS transistor will be interpreted as a logic 1. The programming of the memory (placement of the thin-gate oxide transistors) is performed during the manufacturing process.

static versus dynamic—Aside from the organization of the ROM, which defines its bit capacity, the most important parameter in most applications is probably access time. *Access time* is defined as the time required for a valid output to appear after a valid input has been applied.

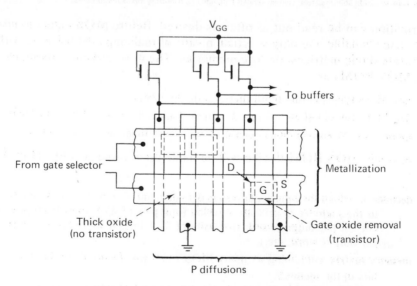

Fig. 5–48 Physical arrangement of MOS/LSI ROM memory matrix. (Courtesy Texas Instruments)

In a *static* ROM there are no clocks required. If a valid input address is applied to the memory, after the expiration of the required access time, a valid output will appear. The output will remain valid as long as the input address remains unchanged.

In a *conventional dynamic* ROM, the information is clocked in and clocked out. The output will remain valid only for a certain period. Dynamic ROMs are advantageous in implementing synchronous logic. To take advantage of their logic flexibility, and to allow the output to be kept valid as long as desired, Texas Instruments has designed latches on the outputs of all dynamic ROMs. Also Texas Instruments ROMs do not require clock drivers since these have been incorporated on the chip.

typical applications—ROMs are well suited to such applications as: *look-up tables,* where the output is a mathematical function of the input; *code conversion,* where the input is one code (e.g., EBCDIC) and the output is another code (USACII); *character generator,* where an alphanumeric character is represented by a binary word; and *random logic,* where the ROM is used to perform Boolean algebra (programmed to provide outputs that are Boolean functions of the input variables).

5–3.3.3 *Programmable Logic Arrays*

A programmable logic array (PLA) is essentially a large ROM with a nonexhaustive decode section adapted to the implementation of random logic. The term random logic, as applied here, means a logic circuit that is not strongly structured, as opposed to circuits such as shift registers, ROMs, etc. When a random logic circuit is implemented by the usual custom MOS process, a large part of the chip area is used for interconnection between basic cells, and is often wasted. The PLA approach minimizes this waste by providing the designer with a programmable circuit already fabricated on a chip (making the best use of space).

In order to program the PLA, only one photomask is modified, and this is accomplished easily and economically. The photomask programs the matrices of the PLA, in accordance with logic equations written by the designer. From a user's standpoint, logic design with PLAs is easy. With this approach, the designer writes down the logic equations of each output in terms of external inputs and feedback inputs. Once this is done, the programming of the matrices (modification of the photomask) is handled by a computer. A *software package* bulletin describes in detail the mechanical aspects of the operation; therefore, such details will not be repeated here.

In effect, from a user's standpoint, the PLA bridges the gap between a custom MOS logic array, and an off-the-shelf ROM.

5-3.3.4 *Random-Access Memories*

MOS random-access memories (RAMs) are ideally suited for those applications requiring high speeds and low power dissipations. MOS RAMs can replace core memories used for scratch pads or for computers, while offering faster access times and a simpler drive circuitry.

RAMs presently on the market are either static or dynamic (see Fig. 5-49).

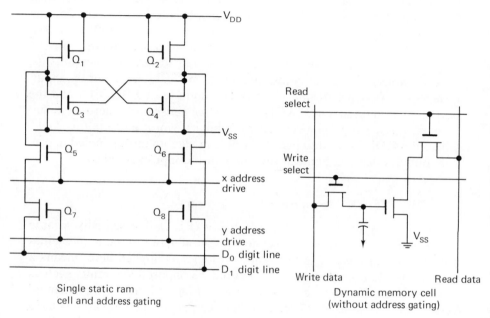

Single static ram
cell and address gating

Dynamic memory cell
(without address gating)

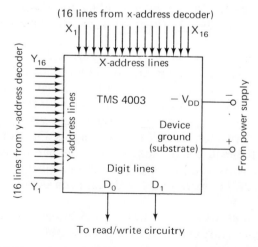

Fig. 5-49 MOS/LSI random-access memory (RAM) arrangement. (Courtesy Texas Instruments)

In the static RAM, a MOS flip-flop is used to store the information. Clocks are not needed. The data will stay in storage as long as the power is maintained.

In the dynamic RAM, MOS capacitors are used as data storage elements. Data must be refreshed to ensure integrity.

RAMs can also be defined as:

Fully-decoded memories, where a binary address determines the location in which the write or read operation is performed;

Two-dimensional decode, where the address of the word is given by an X–Y select.

The static RAM cell of Fig. 5–49 is an example of two-dimensional decode. Assume that the cell is used in a memory with 16 X-address and 16 Y-address lines. Any one of the 256 bits (cells) can be selected by driving one X- and one Y-address line in coincidence. The cell is driven by the X-line through Q_5 and Q_6, and by the Y-line through Q_7 and Q_8.

The state of the flip-flop (Q_1 through Q_4) is sensed by observing outputs D_0 and D_1. The same D_0 and D_1 digit lines are used to write information into an addressed bit.

5–3.3.5 *Content-Addressable Memory*

A content-addressable memory (CAM) is a cell in which all the words contained can be matched simultaneously against an argument word, and outputs can be given wherever a *true match* is obtained. The CAM concept has been around for many years, but cost has made the technique impractical. With improved MOS technology, the CAM has become a working tool for logic design.

Figure 5–50 shows the functions of a CAM, as well as a single CAM cell. Each word has a Write input which can also be used to interrogate that word for a match. Two bit lines run through each column of cells allowing the word data to be written in. The cells may also be used for reading the contents of a word, or for masking parts of the argument word where an irrelevant (don't care) state exists.

In the basic CAM cell, transistors Q_1, Q_2, Q_9 and Q_{10} compose a flip-flop for data storage. Transistors Q_7 and Q_8 are selection transistors which, when turned on by the application of a negative voltage to the W (write) line, connect the flip-flop to the B_0 and B_1 bit lines. When the W line is at a 0, Q_7 and Q_8 are off, isolating the flip-flop from the bit lines.

Transistors Q_3 through Q_6 and Q_{11} perform the Interrogation logic. For this mode, each W line is grounded through a small resistor and Q_{11} is turned on. Transistors Q_3 through Q_6 compare the state of the memory cell flip-flop with the voltages externally applied to the bit lines. Thus, the word lines are controls for Write and Read operations, but are outputs for the Interrogate operation. The bit lines are inputs for Write and Interrogate, but outputs for Read.

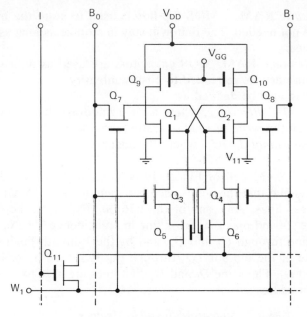

Single cam cell

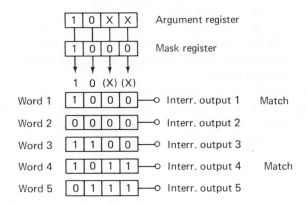

Function of content-addressable memory

Fig. 5–50 MOS/LSI content-addressable memory (CAM) arrangement. (Courtesy Texas Instruments)

5–3.3.6 *Buffers and Analog Switches*

The MOS/LSI line also includes a group of buffers and analog switches. These devices are essentially the same as corresponding two-junction logic devices, except that MOS transistors are involved. Therefore, we will not go into any great detail concerning circuit opera-

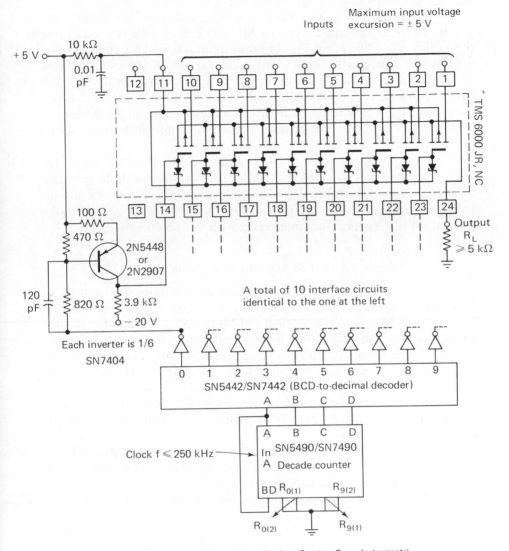

Fig. 5–51 Typical MOS/LSI 10-channel analog switch application. (Courtesy Texas Instruments)

tion. However, since the interface and noise filtering action is somewhat different for MOS, these subjects require some discussion.

A typical 10-channel analog switch application is shown in Fig. 5–51. The switch is used in a direct-coupled multiplexer addressed from TTL devices. In this circuit, each input is connected sequentially through a MOS switch to an output circuit represented by load resistance R_L. A TTL counter and decimal decoder are used to obtain sequential driving from a single clock.

An interface circuit using a PNP transistor translates TTL output voltage levels to those required by the MOS switch. A +5V is used to turn the switch off; −20V is used to turn the circuit on. Because the transistor saturates, a storage time exists which delays turn-off. This delay (about 150 to 300 nS) is used in the interface circuit shown to allow the previous MOS switch to turn-off completely before the next one turns-on. Clock frequency, t_{clock}, is limited by interface circuit storage and fall times to about 250 kHz, before these times become an appreciable fraction of a clock cycle.

The substrate is biased at +12V to allow the drains and sources of the MOS switches to go positive without forward-biasing the drain-substrate and source-substrate diffused diodes. Only leakage current flows into the substrate, so the simple RC filter shown is sufficient to prevent noise from the +12V supply from interfering with switch operation.

5–3.3.7 *MOS/LSI System Compatibility*

Most designs presently under consideration use both MOS/ LSI and two-junction technologies in order to take full advantage of the low-cost and high-packaging density of MOS/LSI, as well as the flexibility of two-junction techniques for low-complexity functions. The interface of different logic families requires that the circuits operate at a *common-supply voltage,* and have *logic-level compatibility.* In addition, the devices must maintain *safe power dissipation levels,* and good *noise immunity* over the operating temperature range. The following notes apply to the MOS/LSI line of Texas Instruments. The subject of interfacing is discussed further in Sec. 5–4.

power supplies—Two manufacturing technologies are common in MOS/ LSI, and common throughout the industry: high-threshold and low-threshold MOS. The power supply requirements are:

	V_{SS}	V_{DD}	V_{GG}
High Threshold	0	−12V	−24V
Low Threshold	0	−5V	−17V

where

V_{SS} is the substrate supply

V_{DD} is the drain supply

V_{GG} is the gate supply

The drain supply will draw most of the current. Some circuits are designed to use only one power supply (saturated logic). V_{DD} and V_{GG} are then common.

To use MOS in a system, it is often convenient to translate all of the power supply voltage to a certain voltage. The common arrangement is:

	V_{SS}	V_{DD}	V_{GG}
High Threshold	+12V	0V	−12V
Low Threshold	+5V	0V	−12V

Some high-threshold devices are specified at $V_{GG} = -28V$ and $V_{DD} = -14V$.

input compatibility—Referencing all voltages to V_{SS}, the input swing on most MOS circuits is as follows:

	High Level	Low Level
High Threshold	0 to −3V	−9V to −24V
Low Threshold	0 to −1.5V	−4.2 to −17V

In relation to the translated power supplies, the input swing becomes:

	High Threshold	Low Threshold
V_{SS}	+12V	+5V
V_{DD}	0V	0V
V_{GG}	−12V	−12V
High Level	+9V to +12V	+3.5V to +5V
Low Level	+3V to −12V	0.8V to −12V

In all cases, the input of the MOS circuit will look like a very-high impedance, and input compatibility is achieved by the circuits of Fig. 5–52. The value of resistor R varies, depending on speed-power requirements. In many cases, this resistor R is diffused on the MOS chip. For low-threshold MOS, the resistor assures that the worst-case TTL output is pulled up to at least 3.5V for proper MOS circuit operation.

output compatibility—Three types of buffers are commonly used on MOS devices: *open-drain, internal pull-up,* and *push-pull.* These buffer arrangements are shown in Fig. 5–53.

With open-drain and internal pull-up, the buffer is simply a current switch. In the OFF state, the impedance of the buffer is extremely large, while in the ON state impedance is typically under 1K. A discrete resistor or a MOS transistor may be used as a load with an *open-drain buffer.*

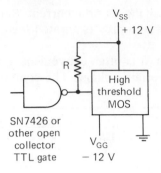

R = see text

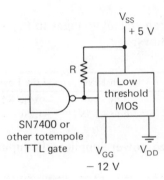

Fig. 5-52 MOS/LSI system imput compatibility. (Courtesy Texas Instruments)

This resistor or the transistor may be internal to the MOS circuit. When the load transistor is internal to the MOS, the buffer is called an *internal pull-up buffer.*

If the MOS is high-threshold with an open-drain buffer, the output can be made compatible with TTL, as illustrated in Fig. 5–54. Resistor R_2 provides the necessary current sink for the TTL input. Resistor R_1 limits the positive excursion to +5V.

If the MOS is low-threshold, V_{SS} is translated up to +5V instead of to +12V, eliminating the need for R_1. Further, if R_2 is on the chip (in resistor form or a MOS load transistor), no external components are necessary, permitting direct coupling of the MOS output to the TTL input, as shown in Fig. 5–55.

There are *two common types of push-pull buffer,* as seen in Fig. 5–56. The unsaturated push-pull buffer is the most commonly used for low-threshold circuits, and permits direct TTL compatibility without external components (as well as direct compatibility with other low-threshold MOS circuits).

clocks—MOS clock requirements depend on the circuit, as is the case with other logic families. For example, no clocks are required for static

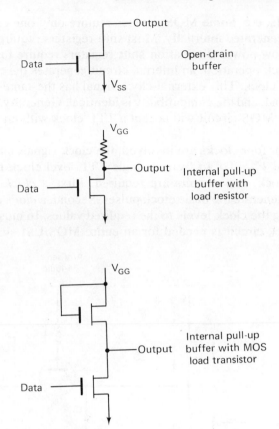

Output

Data —

V_{SS}

Open-drain
buffer

V_{GG}

Output

Data —

Internal pull-up
buffer with
load resistor

V_{GG}

Output

Data —

Internal pull-up
buffer with MOS
load transistor

Fig. 5–53 MOS/LSI system output compatibility. (Courtesy Texas Instruments)

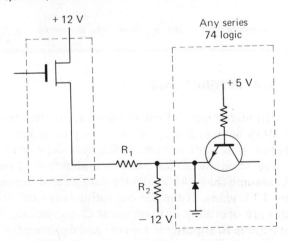

+ 12 V

Any series
74 logic

+ 5 V

R_1

R_2

– 12 V

Fig. 5–54 MOS/LSI high-threshold (with an open-drain buffer) to TTL interface. (Courtesy Texas Instruments)

RAMs, ROMs, etc. Some MOS devices require only one clock, with all other clocks generated internally. Most shift registers require two clocks. High-speed, low-power dissipation shift registers require four clocks.

For one clock operation, an internal circuit generates the clocks from a single outside clock. This external-clock signal has the same swing as the data input signal, and the compatibility is identical. Generally, single-clock low-threshold MOS circuit will accept a TTL clock without adding components.

When two or four clocks are involved, the clock signals must swing between V_{SS} and V_{GG}. To go from a single TTL-level clock to a multiple MOS-level clock, two circuits are required. First, a *clock generator* is necessary to generate the basic clock pulses. Second, a *clock driver* is necessary to bring the clock levels to the required values. In most cases, only one basic clock circuit is needed for an entire MOS/LSI system.

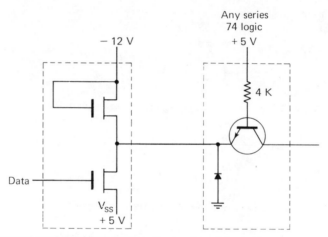

Fig. 5-55 MOS/LSI low-threshold to TTL interface. (Courtesy Texas Instruments)

5-4. INTERFACING

No matter which load and drive characteristics are involved for a digital MOS device, it may be necessary to include some form of interfacing between devices to provide the necessary drive current. This is especially true when the device datasheets show a very close tolerance. For example, assume that a MOS device with a rated fan-out of 3 is used to drive three TTL gates. If the fan-out rating is typical or average, and the three gates are operating in their worst case condition, the MOS device may not be able to supply (or *source*) and dissipate (or *sink*) the necessary current.

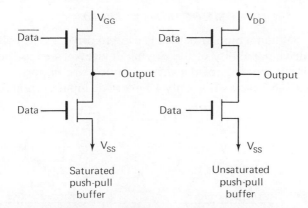

Fig. 5–56 Two common types of push-pull buffers used with MOS/LSI. (Courtesy Texas Instruments)

It is not practical to have a *universal* interfacing circuit for the MOS devices of all manufacturers. Likewise, it is not even possible to have a universal circuit for interface between the MOS device of one manufacturer and all other logic families. For one thing, the logic voltage levels (for 0 and 1), the supply voltages, and the temperature ranges vary with MOS manufacturers, and with logic family. For that reason, most MOS manufacturers publish interfacing data for their particular lines.

We will make no attempt to duplicate all of this data here, but instead, concentrate on the interface requirements for RCA COS/MOS devices. A careful study of this information should provide the reader with sufficient background to understand the basic problems involved with interfacing MOS logic systems.

5–4.1 COS/MOS–TTL/DTL Interface

When interfacing one type of digital IC with another, attention must be given to the logic swing, output drive capability, dc input current, noise immunity, and speed of each type. Figure 5–57 shows a comparison of these parameters for COS/MOS, medium power TTL, and medium power DTL. The supply voltage column of Fig. 5–57 shows that both saturated bipolar (two-junction) and COS/MOS devices may be operated at a supply voltage of 5V. Both logic forms are directly compatible at this supply voltage (with certain restrictions).

Figure 5–57 also shows the voltage characteristics required at the output and input terminals of saturated logic devices, as well as the COS/MOS input and output characteristics at $V_{DD} = 5$V. The COS/MOS devices are designed to switch at a voltage level of one-half the power supply voltage. However, TTL/DTL devices are designed to switch at about $+1.5$V, which is not one-half of the supply voltage.

5-4.1.1 *COS/MOS Driven by Bipolar*

When a bipolar device is used to drive a COS/MOS device, the output drive capability of the driving device, as well as the switching levels and input currents of the driven device, are important considerations. Figure 5–57 shows that only 10 pA of dc input current are required

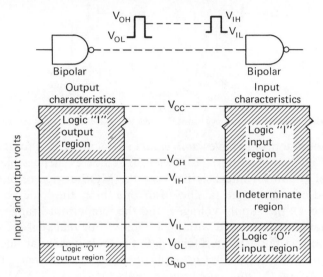

Interface voltage characteristics required at the output and input terminals of saturated logic devices.

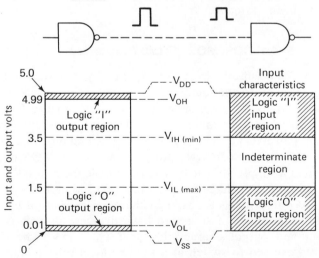

COS/MOS input and output characteristics at a power-supply voltage of 5 volts

Fig. 5–57 COS/MOS to DTL/TTL interface characteristics. (Courtesy RCA)

by a COS/MOS device in either the 1 or 0 state. The input thresholds (for the driven COS/MOS device) are 1.5 and 3.5V. Thus, the output of the TTL/DTL driver must be no more than 1.5V (0 logic voltage) and no less than 3.5V (1 logic voltage) in order to obtain some noise immunity.

current sinking—Figure 5–58 illustrates the low state operation of a loaded bipolar driver stage. When the output drive circuit of the bipolar stage is in the low state, the collector is essentially at ground potential. The ON transistor must go into saturation in order to assure a reliable logic 0 level (0 to 0.4V). To attain this voltage level, there should be a

Family	Supply voltage (volts)	Logic swing/output drive capability	DC input current	Noise immunity	Propagation delays
COS/MOS	3.0 to 15	V_{SS} to V_{DD} (driving COS/MOS) Output drive is type dependent (see text)	10 pA (typical) 1 and 0 state	1.49 at $V_{DD} = 5$ V The switching point occurs from 30% to 70% of V_{DD} which is 1.5 V to 3.5 V at $V_{DD} = 5$ V	35 ns (typical) for inverter $C_L = 15$ pF
DTL and TTL	5	0 state: 0.4 V max. at $I_{sink} = 16$ mA 1 state: 2.4 V min. at $I_{load} = -400$ μA	0 state: -1.6 mA max. 1 state: 40 μA max.	at $V_{CC} = 5$ V 0.4 V guaranteed The switching point occurs from 0.8 V to 2 V	20 ns (typical) for inverter $C_L = 15$ pF

Comparison of COS/MOS, TTL, DTL interfacing parameters

Logic voltage	Description	Voltage (volts)
V_{OL}	Maximum output level in low-level output state	0.4
V_{OH}	Minimum output level in high-level output state	2.4
V_{IL}	Maximum input level in low-level input state	0.8
V_{IH}	Minimum input level in high-level input state	2.0
V_{CC}	Positive supply voltage	5.0 ± 0.5
	Operating temperature range: $-55°$ to $+125°$C—full temperature range product $0°$ to $+85°$C—limited temperature range product	

Common logic voltages, supply voltage, and operating temperature range required to interface with DTL/TTL circuits

Fig. 5–57 (Cont.)

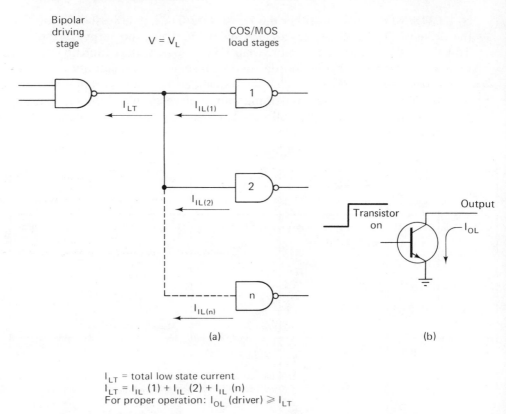

I_{LT} = total low state current
$I_{LT} = I_{IL} (1) + I_{IL} (2) + I_{IL} (n)$
For proper operation: I_{OL} (driver) $\geqslant I_{LT}$

Fig. 5–58 (a) Low-state operation of a loaded bipolar driver stage; (b) typical bipolar output-drive circuit in the low state.

high impedance path from the output to the power supply. Current sinking capability is not a problem in this configuration because the COS/MOS devices have extremely high input impedances (typically 10^{11} ohms). Neither is the voltage level a problem; the COS/MOS devices have high noise immunity (greater than 1V).

current sourcing—Current flows from the V_{CC} terminal of a saturated logic (two-junction) output device into the input stages of the load. That is, the output device acts as the current source for the load. Figure 5–59 shows high-state operation of a loaded bipolar driver stage. Whenever a typical bipolar driver circuit is in the high state, a pull-up configuration (resistor or transistor) ties V_{CC} to the output pin. The total load configuration should not draw sufficient current to reduce the output voltage level below the V_{IH} required by the COS/MOS devices. (V_{IH} is the maximum acceptable input level for the device in the high-level input state, as shown in Fig. 5–60).

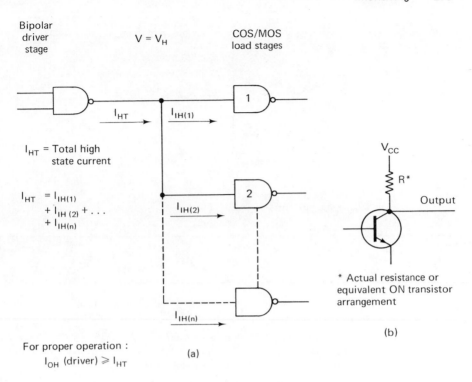

Bipolar driver stage

$V = V_H$

COS/MOS load stages

I_{HT} = Total high state current

I_{HT} = $I_{IH(1)}$ + $I_{IH(2)}$ + ... + $I_{IH(n)}$

I_{HT}

$I_{IH(1)}$

$I_{IH(2)}$

$I_{IH(n)}$

1

2

V_{CC}

R^*

Output

* Actual resistance or equivalent ON transistor arrangement

(b)

For proper operation :
  I_{OH} (driver) $\geqslant$ I_{HT}

(a)

I_{OH} = Maximum permissible output driver current in high state ("1") (driver leakage)
I_{OL} = Maximum output driver sinking current in low state ("0")
I_{IL} = Low state input current drawn from the load stage (to the driver)
I_{IH} = High state input current flowing into the load stage from driver

Fig. 5–59 (a) High-state operating of a loaded bipolar driver stage; (b) typical bipolar output-drive circuit in the high state.

There are three bipolar output configurations to consider: *resistor pull-up, open collector,* and *active pull-up.*

resistor pull-up—Devices with resistor pull-ups, as shown in Fig. 5–59, present no problem in the interface with COS/MOS devices.

open collector—Devices with open collectors require an *external pull-up* resistor, as illustrated in Fig. 5–61. The selection of the external pull-up resistor requires consideration of fan-out, maximum allowable collector current in the low state ($I_{OL}max.$), collector-emitter leakage current in the high state (I_{CEX}), power consumption, power-supply voltage and propagation delay times.

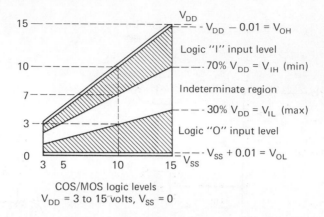

COS/MOS logic levels
V_{DD} = 3 to 15 volts, V_{SS} = 0

Logic voltage symbol	Description
V_{OH}	Minimum guaranteed noise free output level of device in high-level output state
V_{IH}	Minimum acceptable input level for device in high-level input state
V_{IL}	Maximum acceptable input level for device in low-level input state
V_{OL}	Maximum guaranteed noise free output level of device in low-level output state
V_{NL}	Maximum (positive) noise level tolerated at low level state
V_{NH}	Maximum (negative) noise level tolerated at high level state
$V_{OL} + V_{NL} \geqslant V_{IL}$	
$V_{OH} + V_{NH} \leqslant V_{IH}$	
	Operating temperature range
V_{DD}	Positive supply voltage — 55°C to + 125°C (full temperature prod).
V_{SS}	Negative supply voltage — 40°C to + 85°C (limited temperature prod).

Fig. 5–60 Common logic voltages, supply voltage and temperature range for COS/MOS devices. (Courtesy RCA)

The equations of Fig. 5–62 provide guidelines for selection of pull-up resistor maximum and minimum values. These equations neglect the values of I_{IH} and I_{IL} for the MOS devices because such values (typically 10 pA) are insignificant when compared with the value of the bipolar currents.

Assume that these equations are used to find a value of R_X where the conditions are:

$$V_{CC} = 5V \pm 0.5V \qquad V_{IH} \text{ (for the MOS)} = 3.5V$$
$$V_{DD} = 5V \qquad\qquad\qquad I_{OL} = 20 \text{ mA}$$
$$V_{OL} = 0.4V \text{ max} \qquad\qquad I_{CEX} = 100 \text{ } \mu A \text{ max}$$

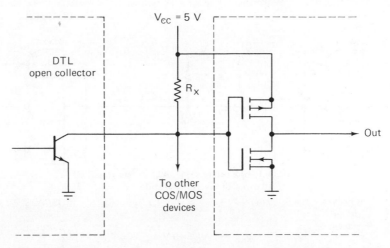

Fig. 5–61 Example of DTL/TTL circuit with open collectors that require a resistor between the output and V_{CC}. (Courtesy RCA)

For one driver (bipolar) and one load (MOS), the values of R_X are:

$$R_{X(min)} = \frac{(5.0 - 0.4)}{0.020} = 230 \text{ ohms}$$

$$R_{X(max)} = \frac{(4.5 - 3.5)}{0.0001} = 10K$$

If more than one driver is used with the open collector drive arrangement, the values of I_{OL} and I_{CEX} must be increased accordingly (I_{OL} and I_{CEX} multiplied by the number of drivers). However, the values of R_X will not be affected by an increase in the number of MOS loads. Of course, if an infinite number of MOS loads are added so that I_{IL} and I_{IH} become significant compared to I_{OL} (for the minimum value of R_X) or to I_{CEX} (for the maximum value of R_X), then the MOS input currents must be included in the calculations as follows:

$$R_{X(min)} = \frac{V_{DD} - V_{OL(max)}}{I_{OL} - I_{IL}(MOS)}$$

$$R_{X(max)} = \frac{V_{CC(min)} - V_{IH(min)}(MOS)}{I_{CEX} + I_{IH}}$$

For short propagation delay times with an external pull-up resistor, it is best to keep R_X small. That is, use the $R_{X(min)}$ value, or the next higher standard value. However, power consumption increases rapidly at values below about 1K so some compromise is usually necessary. Of course,

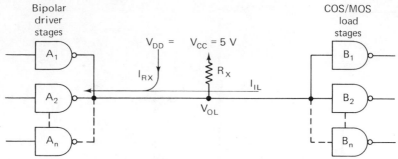

Conditions: Only one driver stage (A$_2$) in low state "O".
All other drivers in high state "I".

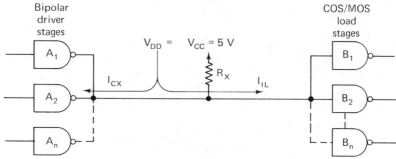

Conditions: All driver stages in high state "I".

$$R_X \text{ (min)} = \frac{V_{DD} - V_{OL \text{ (max)}}}{I_{OL}}$$

$$R_X \text{ (max)} = \frac{V_{CC \text{ (min)}} - V_{IH}^* \text{ (min)}}{(N) \, I_{CEX \text{ (max)}}}$$

*V$_{IH}$ is the value for the COS/MOS device

Fig. 5–62 Bipolar output (with pull-up resistor) driving COS/MOS in low-state and high-state operation. (Courtesy RCA)

final selection of the pull-up resistor depends on what is most important for the intended application: high speed or low power.

Figure 5–63 shows typical speed/power relationships as a function of R_X for two popular bipolar open collector drivers and illustrates the trade-off between speed and power. The power figure shown is the power dissipated in both the bipolar driver and the pull-up resistor.

active pull-ups—When an active pull-up is used, such as the transistor-plus-diode arrangement shown in Fig. 5–64, there can be a problem in the 1 state. This is because the minimum output level (2.4V) cannot assure an acceptable 1 state input for the COS/MOS device. For example, assume that the 2.4V minimum TTL/DTL output level is specified for a

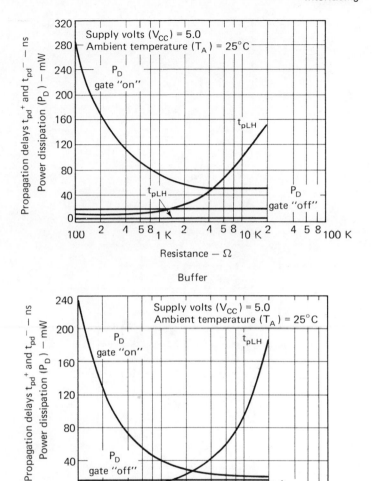

Fig. 5–63 Typical speed-power trade-off of open collector TTL buffer or gate, and pull-up resistor. (Courtesy RCA)

load current of 400 μA (which is typical). This would require approximately 40 MOS devices (with a typical input current of 10 pA).

If only one, or a few, MOS devices are used as a load, the minimum TTL/DTL high-output level will rise to about 3.4 to 3.6V. There is no noise immunity in such a configuration. Therefore, it is recommended that a pull-up resistor be added to V_{CC} from the output terminal of the bipolar device. The selection of this resistor should be based on the calculations of Fig. 5–62 (as previously discussed).

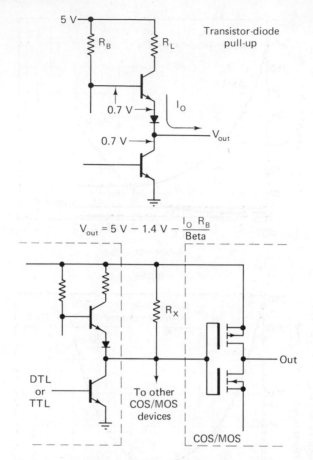

Fig. 5–64 Active pull-up (transistor-diode) to COS/MOS interface. (Courtesy RCA)

When driving a COS/MOS device from an output arrangement as in Fig. 5–64, the driver should not fan-out to TTL/DTL circuits; but only to other COS/MOS devices. *This is because it is not accepted practice to tie devices with active pull-ups together.*

5–4.1.2 *COS/MOS Driving Bipolar*

Figure 5–65 shows COS/MOS devices driving bipolar devices. The current sinking capacity of the COS/MOS device must be considered when the device is driving a medium-power DTL and TTL circuit. Figure 5–57 illustrates that the TTL/DTL device requires no more than 1.6 mA in the 0 input state, and a maximum of 40 μA in the 1 input state.

The COS/MOS device must be capable of sinking and sourcing these currents while maintaining voltage output levels required by the TTL/

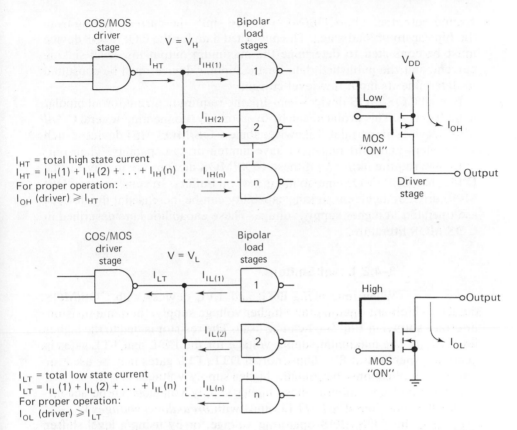

Fig. 5–65 Logic diagram for a COS/MOS device driving a bipolar device, low-state and high-state. (Courtesy RCA)

DTL gate. Any given TTL/DTL gate will switch state at a voltage that ranges from 0.8 to 2V. Hence, the output drive capability of the COS/MOS driver must be at least 40 μA for a given 1-state output voltage of 2V, and at least 1.6 mA for a given 0-state output voltage of 0.8V. In order to provide a noise margin of 400 mV for the driven bipolar device, the COS/MOS device must sink 1.6 mA at a 0 logic state voltage of 0.4V, and 40 μA at a logic 1 level of 2.4V.

current sourcing – In the high-state operation (Fig. 5–65), V_{DD} is normally connected to the driver output through one or more ON P-channel devices which must be able to source the *total leakage current* of the bipolar load stages. The published information for the particular COS/MOS and bipolar devices must be consulted in order to determine the leakage currents (for the logic 1 state), and drive fan-out to be used in the equations shown in Fig. 5–65. Saturated logic devices will not achieve their required switching levels unless this equation has been satisfied.

current sinking – When the output of a COS/MOS driver is in the low-state, an N-channel device is ON and the output is approximately at ground potential. The COS/MOS device sinks the current flowing from the bipolar input-load stage. The published data for the COS/MOS device must be consulted to determine the maximum output low-level sinking current, and the published data for the bipolar device must be consulted to determine its input low-level current.

Not all COS/MOS devices can sink the required current for all bipolar logic families. This problem can be overcome by connecting several COS/MOS devices in parallel. Likewise, some COS/MOS MSI devices (such as counters and shift registers) have limited drive capability. Their outputs may require *buffering* if these COS/MOS devices are to drive TTL/DTL. The COS/MOS line contains several buffers. In some cases, COS/MOS drive and current sinking capability can be increased if the devices are operated at higher supply voltage. These capabilities are described in COS/MOS literature.

5–4.2 Level Shifters

When interfacing DTL and TTL devices with COS/MOS devices which are operated at a higher voltage supply, the same resistor-interface shown in Fig. 5–61 can be used. The resistor is tied to the higher level (V_{DD}). The maximum supply voltage for the DTL and TTL gates is generally specified at 8V. Thus, not all DTL/TTL gates may be used for interface applications that require higher supply voltages (V_{DD}).

Guaranteed operation at these higher supply voltages can be accomplished by selection of DTL/TTL units with *breakdown voltages* $V_{(BR)CER}$ *exceeding* the COS/MOS operating voltage, or by using a level shifter circuit (also known as a *level translator*)(shown in Fig. 5–66). This circuit converts DTL, TTL, and even RTL (resistor-transistor logic) input logic levels to voltages compatible with COS/MOS circuitry. In interface applications, the supply voltage for the translator should be equal to the supply voltage required by the COS/MOS circuitry.

The speed consideration is most important when a separate interface circuit is used. It is desirable (unless high ac noise immunity is a prime consideration) for the speed of the interfacing circuit to be maximum (or at least no slower) than either type of logic joined by the interface. No interfacing device other than a pull-up resistor is required, however, between the COS/MOS and TTL logic at a supply voltage of 5V. Speeds involved when COS/MOS drives TTL (which can be found in the published data for COS/MOS devices) are comparable to the COS/MOS propagation delays. Speeds involved when COS/MOS is driven by TTL, even with a large external resistor, are no slower than delay times for COS/MOS logic circuits. As a result, speed is not a problem in COS/MOS–TTL logic interfacing, provided *clock rates are within the* COS/MOS *range*.

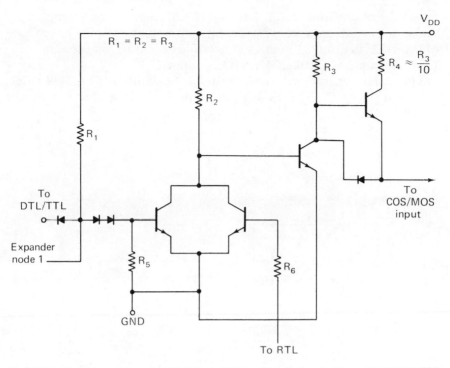

Fig. 5–66 Level translator used to convert DTL, TTL, and RDL input logic levels and voltages compatible with COS/MOS circuitry. (Courtesy RCA)

5–4.3 COS/MOS–HTL Interface

HTL (high threshold logic) circuits operate at voltage levels between 14 and 16V. COS/MOS logic circuits can operate at these voltages as well, but generally are limited to voltages no higher than 15V. HTL circuits are similar to DTL. The major difference is a base-emitter junction in the reverse direction added on the emitter side of the input transistor. This junction is a zener diode which gives a higher threshold switching voltage. Conduction does not occur until the junction breaks down at a voltage of approximately 6.7V, providing the HTL with its high noise immunity.

HTL circuits have more limited temperature range, and dissipate much more power than COS/MOS circuits. Thus, care should be exercised when using the combination in extreme temperature environments. Typically, HTL resistance values vary by about 20 percent from one end of their temperature range to the other. In addition, the transistors used in HTL are sensitive to temperature, and are subject to thermal runaway. The V_{OL} level, propagation delay, and noise immunity of HTL circuits vary widely across the temperature range. COS/MOS circuits show almost negligible variation for these same parameters over a temperature range that is approximately 75 percent wider than that of HTL. In COS/MOS–HTL interface, the main concern in regard to temperature is the HTL parameters, not the COS/MOS parameter.

The same general rules described for COS/MOS–TTL/DTL (Sec. 5–4.1) apply to COS/MOS–HTL. Figure 5–67 illustrates the voltage

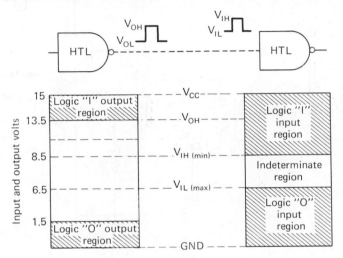

HTL output and input voltage characteristics
at a V_{CC} of 15 volts

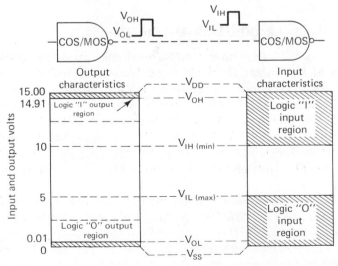

COS/MOS output and input voltage characteristics
at a V_{CC} of 15 volts

Fig. 5–67 COS/MOS and HTL interface characteristics. (Courtesy RCA)

HTL common logic voltages, supply voltage
and operating temperature ranges

Logic voltage symbol	Description	Voltage
V_{OL}	Maximum output level in low-level output state	1.5 V
V_{OH}	Minimum output level in high-level output state	13.5 V
V_{IL}	Maximum input level in low-level input state	6.5 V
V_{IH}	Minimum input level in high-level input state	8.5 V
V_{NL}	Worst case positive noise level tolerated at low level state	5.0 V
V_{NH}	Worst case negative noise level tolerated at high level state	5.0 V
V_{CC}	Positive supply voltage	15.0 ± 1 V
Operating temperature range $-30°C$ to $+75°C$		

Fig. 5–67 (Cont.)

characteristics required at the output and input of an HTL device for a $V_{CC} = 15V$, as well as the same characteristics for a COS/MOS device.

The HTL types either have a built-in pull-up resistor (typically about 15K) or an active pull-up. An external pull-up resistor is unnecessary when COS/MOS devices are being driven by HTL. The dc noise immunity in the high state (logic 1) is 3.5V for an active pull-up and 5V for a resistor pull-up circuit.

The published data should be consulted to be sure that the rise and fall times and the pulse widths of the HTL output are compatible with the required pulse width and input rise and fall times of the COS/MOS circuits. The rules for selection of external pull-up resistors are the same as described in Sec. 5–4.1, except that the values are different. For example, a typical HTL will have values such as:

$$V_{IH} = 8.5V \text{ (min)} \qquad I_{IH}max = 2 \ \mu A$$

$$V_{IL} = 6.5V \text{ (max)} \qquad I_{IL}max = 1.2 \text{ mA}$$

5–4.4 COS/MOS–ECL/ECCSL Interface

Figure 5–68 shows the interface of COS/MOS devices with ECL devices. High-speed ECL (emitter-coupled logic) and ECCSL (emitter-coupled current-steering logic) are nonsaturating two-junction logic families. The V_{CC} to V_{EE} voltage range is fixed from a ground level to -5 or $-5.2V$. Logic 1 to logic 0 values are separated by only 0.3 to 0.5V, depending on the particular type of ECL family used. Figure 5–69 shows some typical ECL/ECCSL values. However, since each manufacturer shows different logic levels for a number of ECL families, care should be taken to use only the applicable values taken directly from the published data.

A logic 1 is the most positive frame of reference, and a logic 0 the most

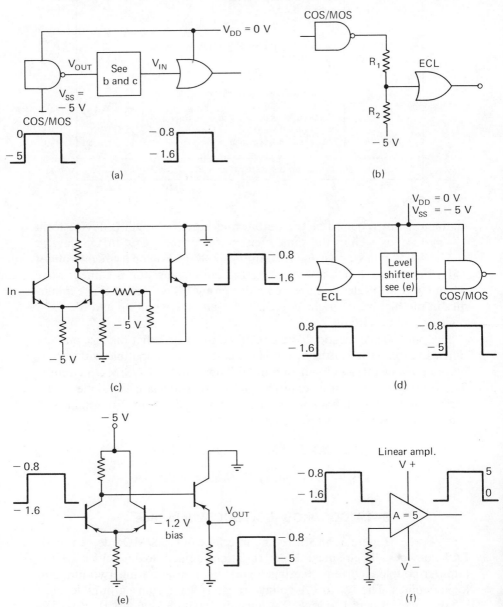

Fig. 5–68 COS/MOS and ECL/ECCSL interface characteristics. (Courtesy RCA)

negative. For example, for positive logic, an RCA type CD2150 OR/ NOR gate is at a logic 1 level when its voltage is −0.8V, and at a logic 0 when its voltage is −1.6V (more negative value).

The interfacing of COS/MOS devices driving ECL devices requires a method to reduce the output voltage swing from 0.3 to 0.9V. This can be accomplished with a precise resistor-divider network arrangement [Fig. 5–68(b)], an emitter-follower [Fig. 5–68(c)], or numerous combinations of resistor, diode and transistor configurations. For example, if the COS/ MOS output for a logic 1 is 5V, and the input for an ECL logic 1 is 1V, the resistor-divider network of Fig. 5–68(b) can be chosen to provide a five-to-one voltage division.

Care must be taken to meet the necessary current sinking requirements of the ECL device, particularly in the 0 logic state. An external circuit may be required for COS/MOS to sink the current from the ECL load. Likewise, the published data must be consulted to be sure that the rise and fall times, and pulse widths, of the ECL/ECCSL are compatible with the COS/MOS.

The interfacing of COS/MOS devices that are being driven by ECL/ ECCSL generally includes an amplifier. Amplification is necessary because the COS/MOS requires a greater voltage swing. That is, the ECL −0.8 to −1.6V swing must be amplified to the COS/MOS 0 to 5V swing. The use of a separate transistor circuit, such as that shown in Fig. 5–68(e), is recommended. Proper biasing of the transistor is essential. It is suggested that the V_{SS} level for the COS/MOS circuit be the same as the V_{EE} level of the ECL circuit. This will minimize the number of power supplies as well as provide better interface conditions.

Logic voltage symbol	Description	Voltage range*** From	To
V_{OL}	Maximum output level low-level state	− 1.6	− 1.45*
V_{OH}	Minimum output level high-level state	− 0.8	− 0.795*
V_{IL}	Minimum input level low-level state	− 1.4	− 1.7*
V_{IH}	Maximum input level high-level state	− 0.75	− 1.1
V_{NL}	Worst case positive noise level tolerated at low-level state	0.20*	0.35*
V_{NH}	Worst case negative noise level tolerated at high-level state	− 0.235	− 0.305*
V_{CC}	Positive supply voltage	0	0
V_{EE}	Negative supply voltage	− 5.5	− 5.0

Temperature range + 10 to + 60°C

* At T = + 25°C
** These values are representative of the range for several ECL families

Fig. 5–69 ECL common logic voltages, supply voltages, and operating temperature range. (Courtesy RCA)

5–4.5 COS/MOS to MOS Interface

There are a number of MOS devices which function at the same V_{DD} and V_{SS} ranges of COS/MOS. These devices can be interfaced directly, provided V_{DD} and V_{SS} are the same. Direct interface applies to N-channel MOS devices. If P-channel devices are involved, there may be a problem of logic polarity. If the P-MOS device uses negative logic, the COS/MOS positive logic must be converted. The techniques involved for simple transposition from positive-logic to negative-logic are described in the author's *Handbook of Logic Circuits*.

5–5. INTERPRETING IC MOS DATASHEETS

As is the case with discrete MOSFETs, digital IC MOS device datasheets are presented in various formats.

However, there is a general pattern used by most manufacturers. For example, most MOS IC datasheets are divided into four parts. The first part usually provides a logic diagram and/or circuit schematic, plus truth tables, logic equations, general characteristics, and a brief description of the device. The second part is devoted to tests, and shows diagrams of circuits for testing the MOS IC. These two parts are fairly straightforward, and are usually easy to understand.

This is not necessarily true of the remaining two parts (or one large part in some cases) which have such titles as "maximum limits" or "maximum ratings," and "basic characteristics" or "electrical characteristics." These latter parts are generally in the form of charts, tables, or graphs (or combinations of all three), and often contain the real data needed for design with logic MOS ICs.

The terms used by manufacturers are not consistent. Likewise, a manufacturer may use two different terms to describe the same characteristic when different lines are being discussed.

It is impractical to discuss all characteristics found on digital MOS IC datasheets. However, the following notes should help the user interpret the most critical values.

5–5.1 Electrical Characteristics

It is safe to assume that any electrical characteristic listed on the datasheet has been tested by the manufacturer. If the datasheet also specifies the test conditions under which the values are found, the characteristic can be of immediate value to the user.

For example, if leakage current is measured under worst-case conditions (maximum supply voltage and maximum logic input signal), the leakage current shown on the datasheet can be used for design. However, if the same leakage current is measured with no signal input (inputs grounded or open), the leakage current is of little value. The same is true of such factors as output breakdown voltage and maximum power supply current.

To sum up, if an electrical characteristic is represented as being tested under *typical* (or preferably worst-case) operating conditions, it is safe to take that characteristic as a design value. If the characteristic is measured under no signal conditions, it is probably included on the datasheet to show the *relative* merits of the device. When in doubt, test the device as discussed in Chap. 6.

In the case of *quiescent power values* for a complementary MOS device, always use the maximum value, rather than the typical value, if both are given. Also, consider clock times, or logic operating speed, when calculating power. In a complementary MOS system, most of the power is dissipated *during transition*. Thus, if transition is slow due to long clock pulses, average power consumption is increased.

5–5.2 Maximum Ratings

Maximum ratings are values that must never be exceeded in any circumstance. They are not typical operating levels. For example, a maximum rating of 15V for V_{CC} means that if the regulator of the system power supply fails, and the V_{CC} supply moves up from the normal voltage (5 to 10V, typically), the MOS IC will probably not be burned out; but never design the system for a normal V_{CC} of 15V. Allow at least a 10 to 20 percent margin below the maximum, and preferably a 50 percent margin for power supply (voltage and current) limits. Of course, if typical operating levels are given, these can be used even though they are near the maximum ratings.

5–5.3 Drive and Load Characteristics

Generally, the most important characteristics of digital MOS ICs (from the user's standpoint) are those that apply to the output drive capability, and the input load presented by the IC. This applies to both *combinational logic* and *sequential logic*.

Combinational logic involves circuits having outputs that are a direct function only of present inputs, and involve *no memory function* (adders, coders, decoders, etc.). Sequential logic involves circuits which contain at least one memory element (flip-flops, registers, counters, etc.).

No matter what type of logic is involved, *the user must know how many inputs* can be driven from one MOS IC (without amplifiers, buffers, etc.). Equally important, he should know what kind of load is presented by the input of an IC on the output of the previous stage. Fan-out is a simple, but not necessarily accurate term to describe drive and load capabilities of an IC.

A more accurate system is where actual input and output currents are given. There are four terms of particular importance. These are:

Output logic 1—state source current I_{OH}

Output logic 0—state sink current I_{OL}

Input forward current—I_F

Input reverse current—I_R

The main concern is that the datasheet value of I_{OL} (or whatever term is used in its place) must be equal to or greater than the combined I_F value of all gates (or other circuits) connected to the MOS IC output. Likewise, I_{OH} must equal or exceed the total I_R.

Unfortunately, the same condition exists for datasheet load and drive factors as exists for other electrical characteristics; the values are not consistent.

5-6. NOISE IMMUNITY

The immunity of a digital MOS IC logic gate to noise signal is a function with many variables, such as individual chip differences, fan-in and fan-out, stray inductance and capacitance, supply voltage, location of the noise, shape of the noise signal, and temperature. Moreover, the immunity of a MOS system usually differs from that of an individual MOS device.

Because of these many variables, a generalized noise immunity test for MOS logic is not practical. Instead, it is more practical to test for noise immunity under a specific set of conditions. Ideally, this should be under actual or simulated operation conditions. If this is not possible, the next best arrangement is to test under conditions *usually encountered* in a logic system.

For these reasons, all discussions concerning noise immunity are contained in the test section (Chap. 6).

5-7. COS/MOS DEVICES USED AS ASTABLE AND MONOSTABLE OSCILLATORS

It is possible to connect complementary MOS devices in simple circuits to form astable and monostable oscillators (or multivibrators). The following paragraphs describe a few examples. All of the circuits described involve the use of MOS NAND and NOR gates, or MOS inverters. (Inverters are formed when all inputs of a NAND or NOR gate are tied together.)

5-7.1 Astable Oscillators

Figure 5-70 illustrates an astable multivibrator circuit that uses two COS/MOS inverters. This simple circuit requires only one resistor and one capacitor. Operation of the circuit is as follows:

When waveform 1 at the output of inverter B is high (logic 1), the input to inverter A is also high. Under these conditions, the output of inverter A is low, and capacitor C is charged. Resistor R is returned to the output of inverter A to provide a path to ground for discharge of capacitor C.

As long as the output of A is low, the output of inverter B is high. As capacitor C discharges through R, the voltage shown as waveform 2 approaches and passes through the transfer voltage point of inverter A. At the instant of crossover, the output of A becomes high, the output of B becomes low (logic 0), and C charges in the opposite direction.

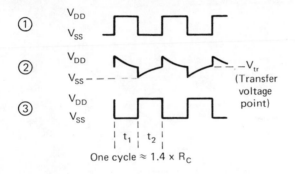

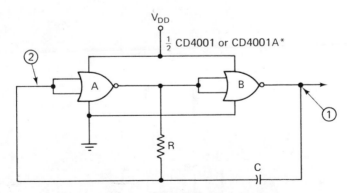

* This circuit can also be implemented by use of other COS/MOS devices such as NAND gates or inverters in place of NOR gates.

Fig. 5-70 Astable multivibrator implemented with two NOR gates connected as inverters. (Courtesy RCA)

The crossover process is repeated at a rate determined by the time constant of R and C. Because of the input-diode protection circuits included in the COS/MOS, the generated drive waveform is clamped between V_{DD} and V_{SS}. Consequently, the time to complete one cycle is *approximately* 1.4 times the RC time constant, because *one time constant* is used to control the switching of *both states* of the multivibrator circuit.

The time constant of the Fig. 5–70 circuit is *voltage dependent,* making the (1.4 × RC) figure an approximation. However, assuming that the transfer voltage point V_{tr} varies as much as 33 to 67 percent of V_{DD}, the time constant multiplication factor will vary from about 1.4 to 1.5 times RC. Thus, the maximum variation in the time period is about 9 percent with a ±33 percent variation in V_{tr} (from unit to unit).

The oscillator can be made independent of supply voltage variations by use of a resistor R_S in series with the input lead to inverter A, as shown in Fig. 5–71. Resistor R_S should be at least twice the value of R_{tc}. This will

$$T = -\ RC\ \ln\left[\frac{V_{tr}}{(V_{DD} + V_{tr})} + \ln\frac{(V_{DD} - V_{tr})}{2\,V_{DD} - V_{tr}}\right]$$

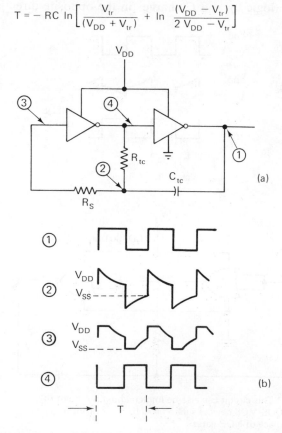

Fig. 5–71 Astable multivibrator (independent of supply voltage) implemented with two COS/MOS inverters. (Courtesy RCA)

allow the voltage waveform generated at the junction of R_S, R_{tc} and C_{tc} to rise to $V_{DD} + V_{tr}$. The waveform is still clamped at the input between V_{DD} and V_{SS}, as illustrated by the waveforms in Fig. 5–71(b).

The use of resistor R_S provides several advantages in the circuit. First, because the RC time constant controls the frequency, the over-all maximum variations in time period are reduced to less than 5 percent with variations in V_{tr}. Resistor R_S also makes the frequency independent of supply-voltage variations.

The time period T for one cycle can be computed using the equations shown in Fig. 5–71. The frequency of the oscillator is the reciprocal of the time period, or $1/T$.

For example, assume that V_{DD} is 10V, V_{tr} is 5V, R is 0.4 megohm, C is 0.001 μF, and R_S is 0.8 megohm (twice the value of R). $RC = (0.4 \times 10^{16})(0.001 \times 10^{-6}) = 0.0004$ second, or 0.4 mS. The time period T is found by:

$$T = 0.4 \ mS \times \left[1. \ \frac{5}{10 + 5} + 1. \ \frac{(10 - 5)}{(20 - 5)}\right]$$

$$= 0.4 \ mS \times 2.66 = 1.064 \ mS$$

The frequency is approximately 1 kHz.

The astable multivibrator of Fig. 5–70 or 5–71 can be gated ON and OFF by use of a NOR or NAND gate as the first inverter. Such an arrangement is shown in Fig. 5–72.

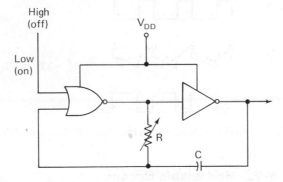

Fig. 5–72 Astable multivibrator in which a NOR, COS/MOS, or NAND gate is used as the first inverter to permit gating of the multivibrator. (Courtesy RCA)

5–7.2 Variable Duty Cycle Astable Oscillators

A true square-wave pulse is obtained only when V_{tr} occurs at the 50 percent point. The duty cycle can be controlled if part of the resistance in the RC time constant is shunted out with a diode, as in Fig. 5–73.

Because adjustment of this diode shunt to obtain a specific pulse duty factor causes the frequency of the circuit to vary, a frequency control R_3 is added for compensation. It may be necessary to reverse the diode to obtain the desired duty factor. The frequency of any of the circuits (Figs. 5–70 through 5–72) can be made variable by use of a potentiometer for resistor R.

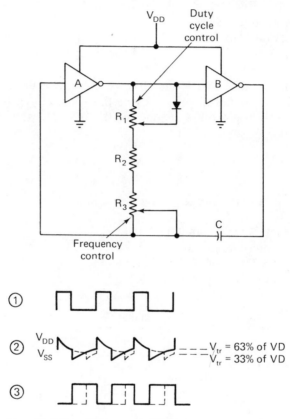

Fig. 5–73 Variable duty cycle astable multivibrator using COS/MOS inverters. (Courtesy RCA)

5–7.3 Monostable Circuits

Figure 5–74 shows a compensated monostable multivibrator type of circuit that can be triggered with a negative-going pulse (V_{DD} to ground). In the quiescent state, the input to inverter A is high, and the output is low. Thus, the output of inverter B is high. When a negative-going pulse or spike is introduced into the circuit as shown in the waveforms, capacitor C_1 becomes negatively charged to ground, and the output of inverter A goes high. Capacitor C_2 then charges to V_{DD} through the diode D_1 and inverter A, and the output of inverter B becomes low.

As capacitor C_1 discharges negatively, it charges to the opposite polarity through R_1 to V_{DD} (waveform 2). The output of inverter A remains high until the voltage waveform generated by the charge of C_1 passes through the transfer voltage point of inverter A. At that instant, the output of inverter A goes low.

Diode D_1 temporarily prevents the discharge of capacitor C_2, which was charged when inverter A was high (waveform 3). Capacitor C_2 then commences to discharge to ground through R_2 (waveform 4). The output of inverter B remains low until the waveform is generated by the discharge of C_2 passing through the transfer voltage point of inverter B. At that point, the output returns to its high state (waveform 5).

When two inverters are fabricated on the same chip their transfer voltage point will be similar. This is an advantage in that any variations in V_{tr} are cancelled out.

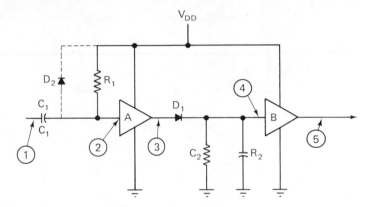

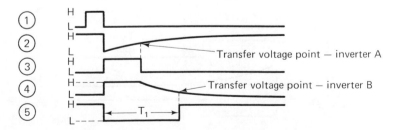

Fig. 5–74 Compensated monostable multivibrator using COS/MOS inverters. (Courtesy RCA)

When $R_1 = R_2$ and $C_1 = C_2$, the period T is *approximately* equal to the RC time constant. However, from a practical standpoint, the period will usually be somewhat greater than the time constant. For example, if $R_1 = R_2 = 1$ megohm, and $C_1 = C_2 = 0.001$ μF, the period will be about 1 to 1.1 mS.

Figures 5–75 and 5–76 show variations of the monostable circuit, together with their associated waveforms. The circuit of Fig. 5–75 triggers on the negative-going excursions of the input pulse, in the same manner as the circuit of Fig. 5–74. The output pulse is positive-going and is taken from the first inverter (waveform 3). No external diode is required.

The circuit of Fig. 5–76 triggers on the positive-going excursion of the input pulse, and then locks back on itself until the RC time constants complete their discharge.

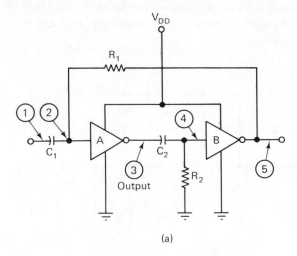

(a)

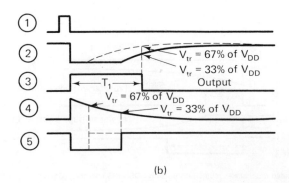

(b)

Fig. 5–75 Monostable multivibrator triggered by a negative-going input pulse. (Courtesy RCA)

Note that the circuits of Fig. 5–75 and 5–76 *cannot* be retriggered until they return to their quiescent states.

5–7.4 Voltage-Controlled Oscillator

Figure 5–77 shows a circuit similar to that of Fig. 5–71. However, in the Fig. 5–77 circuit, C is made variable by C_X, and R is made variable by adjustment of V_A (which is applied to the gate of an N-channel MOS device). The value of R varies from approximately 1 to 10K. These limits are determined by the parallel combination of R_1 (10K) and the N-channel MOS device resistance (which varies from about 1K when fully ON, to about 10^9 ohms when fully OFF).

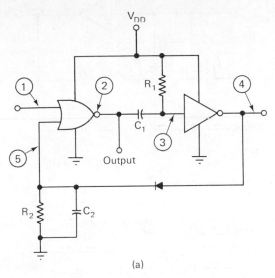

(a)

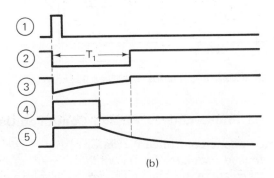

(b)

Fig. 5–76 Monostable multivibrator triggered by a positive-going input pulse. (Courtesy RCA)

When $V_A = V_{SS}$, the N-channel device is OFF, and R = the parallel combination of R_{OFF} and R_1, or about 10K, since R_{OFF} is much greater than R_1.

When $V_A = V_{DD}$, the N-channel device is ON, and R = the parallel combination of R_{ON} and R_1, or about 1K, since R_{ON} is much smaller than R_1.

The oscillator center frequency is varied by adjustment of C_X. The table in Fig. 5–77 shows a comparison of the output waveform period as a function of V_{DD} and V_A.

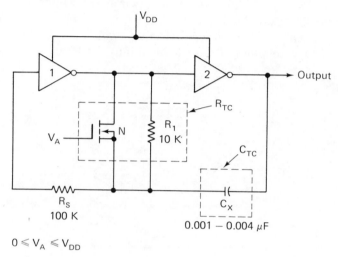

Inverters and N-channel device are available in a single COS/MOS package (CD4007 or CD4007A).

	Period (μS)		
V_A	V_{DD} = 5 V	V_{DD} = 10 V	V_{DD} = 15 V
0	120	54	48
5	115	45	41
10	— —	32	30
15	— —	111	24

Fig. 5–77 Voltage-controlled oscillator using COS/MOS inverters. (Courtesy RCA)

5–7.5 Phase-Locked Voltage-Controlled Oscillator

The voltage-controlled oscillator of Fig. 5–77 can be operated as a phase-locked oscillator by the application of a frequency-controlled voltage to the gate of the N-channel device. Figure 5–78 shows the block diagram of an FM discriminator using the phase-locked voltage-controlled oscillator (VCO).

The VCO block is the same circuit as Fig. 5–77. The output of a standard phase comparator is fed into the gate of the N-channel device (V_A). If the two inputs to the phase comparator are different, the change of V_A causes the output frequency of the VCO to change. This change is divided by 2^N (through a counter), and fed back to the phase comparator.

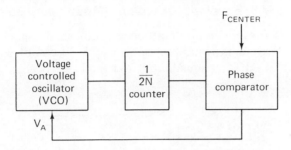

Fig. 5–78 Voltage-controlled-oscillator used as phase-locked loop. (Courtesy RCA)

5–7.6 Voltage-Controlled Pulse-Width Circuit

Figure 5–79 illustrates a further modification of the Fig. 5–71 circuit to provide a modulated pulse width. That is, the pulse width can be controlled by a voltage at V_A. The frequency will not be affected by the V_A voltage, provided the value of R_X is high. The values shown provide pulse periods and pulse widths, as indicated in the table.

Note that the period is determined by component values and V_{DD}, whereas pulse width is primarily dependent on V_A. For example, with the resistance values shown, C at 1500 pF, V_A at 5V, and V_{DD} at 10V, the period is 35 microseconds, and the pulse width is 17.7 microseconds. If V_A is changed to 10V, then pulse width changes to 16.2 microseconds, but period remains at 35 microseconds.

5–7.7 Frequency Multiplier

Figure 5–80 shows a frequency multiplier (doubler) using MOS IC devices. A 2^N multiplier can be realized by cascading the basic circuit with N-1 and other identical circuits. That is, a 2^3 multiplier is formed when three basic circuits are cascaded.

With the basic circuit, the leading edge of the input signal is differentiated by R_1 and C_1, applied to input 1 of the NAND gate. This produces a pulse at the output. The trailing edge of the input pulse, after having been inverted, is differentiated, and applied to input 2 of the NAND gate. This produces a second output pulse from the NAND gate.

In theory, any number of basic circuits can be cascaded. However, the practical limit is between 5 and 10. Also, the RC time constant of R_1C_1 and R_2C_2 sets a limit on the frequency. For best waveforms, the RC time constant should be approximately equal to the period of the pulse signals.

5–7.8 Envelope Detection (Modulation/Demodulation)

Pulse modulation may be accomplished using the circuit of Fig. 5–81. This circuit is a variation of the Fig. 5–71 circuit. The oscillator is gated ON and OFF by signal input 1 at the NAND gate. The number

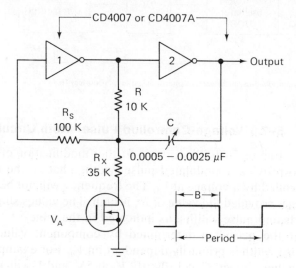

	Pulse width (B)μS		
V_A	V_{DD} = 5 V Period 41.5	V_{DD} = 10 V Period 35	V_{DD} = 15 V Period 33
0	23	19.3	17
5	20	17.7	16.2
10	–	16.2	15.5
15	–	– –	14.3
C = 0.0015 μF			

Fig. 5–79 Voltage-controlled pulse-width circuit using COS/MOS inverters. (Courtesy RCA)

of pulses at the output (waveform 5) during the gated period (waveform 1) is dependent on the RC time constants.

Demodulation or envelope detection of pulse modulated waves is performed by the circuit shown in Fig. 5–82. The carrier burst is inverted by inverter *A*. The first negative transition at point 2, turns on diode *D* to provide a charging path for *C* through the *N*-channel resistance to ground. On the positive transition of the signal at point 2, diode *D* is cut off, and *C* discharges through *R*. The discharge RC time constant is *much greater* than the time of burst duration. Thus, point 3 never reaches the switch point of inverter *B* until the burst has ended.

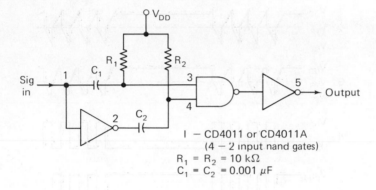

I — CD4011 or CD4011A
(4 — 2 input nand gates)
$R_1 = R_2 = 10 \text{ k}\Omega$
$C_1 = C_2 = 0.001 \text{ }\mu\text{F}$

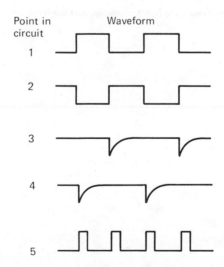

Fig. 5–80 Frequency doubler circuit using COS/MOS devices. (Courtesy RCA)

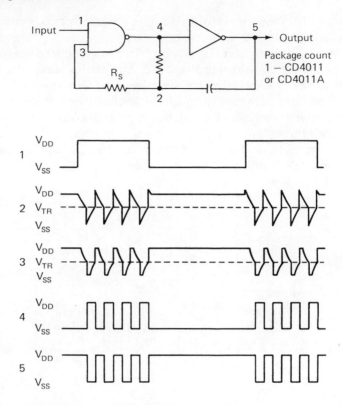

Fig. 5–81 Pulse modulator circuit using COS/MOS devices. (Courtesy RCA)

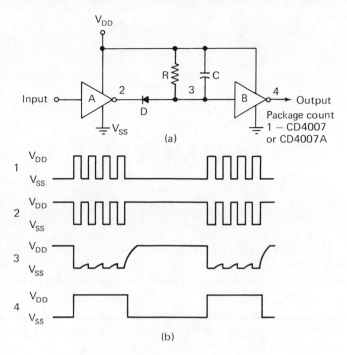

Fig. 5–82 Pulse demodulator circuit using COS/MOS devices. (Courtesy RCA)

6. TESTING MOS DEVICES

This chapter is devoted to test procedures for MOS devices. Keep in mind that MOS devices are used in circuits. Such circuit should be tested according to function. That is, a discrete MOSFET may be used as an amplifier or oscillator. Likewise, logic circuits (gates, registers, counters, etc.) can be fabricated using the complementary MOS principles. The test procedures for MOS circuits, linear and digital, are the same as for two-junction (and vacuum tube) circuits.

Typical circuit test procedures are described in the author's *Manual for Integrated Circuit Users, Handbook of Logic Circuits,* and *Handbook of Practical Electronic Test and Measurements.* These test procedures will not be duplicated here. Instead, we will concentrate on *device tests.*

The tests described in this chapter provide the user with a means of determining the actual or true characteristics of a MOS device. The determination of true characteristics is more important than many users realize. There are several reasons for this.

First, a datasheet may not be available. More likely, the datasheet may not list all of the characteristics needed by the user. Even if all required characteristics are given, the procedures for finding the values may not be given or are not clear.

When test information and characteristics are available in datasheet form, keep in mind that the values given are *typical.* They can vary from device to device, and with *different operating conditions.* There is no substitute for testing each device under actual operating conditions (temperature, power supply, noise, etc.) of the intended use.

Just as important, a MOS device can often be adapted to many uses other than the application intended by the manufacturer. Often, the MOS device manufacturers are surprised at the uses to which their products are adapted. Therefore, they would have no reason for supplying test data (values or procedures) for such applications. Users must devise their own tests and find their own values.

Finally, some MOS device manufacturers assume that all users will automatically know how to test for all characteristics. As a result, they simply omit test data from their literature.

We assume that the reader is thoroughly familiar with basic electronic test equipment. It is particularly important that the reader be familiar with the oscilloscope, voltmeter, signal generator, capacitance meter, pulse generator, R_X meter, and admittance meter. Such equipment is described in the author's *Handbook of Electronic Test Equipment*.

6–1. STATIC TESTS

The following section describes important static (direct-current) tests for MOS devices. All of the characteristics covered in Sec. 1–6.1 are discussed. Keep in mind that MOS devices operate in three modes (type A or depletion only, type B or depletion/enhancement, and type C or enhancement only).

6–1.1 $V_{GS(off)}$ Test

The basic circuit for $V_{GS(off)}$ test is shown in Fig. 6–1. This test applies to depletion only and depletion/enhancement mode devices. As shown, V_{DS} is set at some fixed value, and reverse-bias V_{GS} is adjusted until I_D is at some specific negligible value. Essentially, this is a *cutoff value* test. As an example, the N-channel 3N128 specifies a V_{DS} of 15V, and an I_D of 50 μA. The V_{GS} should be between -0.5V and -8V for the 50 μA of I_D. As a guideline, V_{DS} should be a minimum of 1.5 times $V_{GS(off)}$ to provide proper circuit operation.

6–1.2 $V_{GS(th)}$ or $V_{GS(on)}$ Test

$V_{GS(th)}$ and $V_{GS(on)}$ are essentially the same specification as $V_{GS(off)}$, except that $V_{GS(th)}$ and $V_{GS(on)}$ are generally applied to enhancement only mode devices. The basic circuit of Fig. 6–1 is used, except that V_{GS} is connected for forward bias (gate positive for N-channel). A forward bias is required since I_D flows in an enhancement mode device only when the gate is forward biased. During test, V_{GS} is increased from zero volts until the I_D is at a specified value.

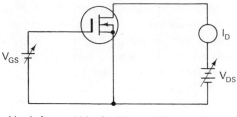

V_{GS} is forward bias for $V_{GS(TH)}$, $V_{GS(ON)}$, $I_{D(ON)}$

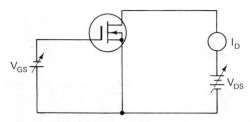

V_{GS} is reverse bias for $V_{GS(OFF)}$, $I_{D(OFF)}$, $V_{(BR)DSX}$

Polarities shown are for N-channel MOS devices.
Reverse all polarities for P-channel

Fig. 6–1 Test circuit for $V_{GS(off)}$, $V_{GS(th)}$, $V_{GS(on)}$, $I_{D(on)}$, $I_{D(off)}$, and $V_{(BR)DSX}$.

6–1.3 $V_{DS(off)}$ and $V_{DS(on)}$ Saturation Test

The basic circuit for saturation tests is shown in Fig. 6–2. Saturation tests are generally of concern only when the MOS device is used for switch or chopper applications. During either test, V_{GS} is held at some fixed value, and V_{DS} is increased until I_D is at maximum, or increases very little for further increases in V_{DS}. Take care not to exceed the maximum V_{DS} when making this test. Generally, $V_{DS(on)}$ is made with some specific value of V_{GS}, whereas $V_{DS(off)}$ is made with V_{GS} at zero (usually shorted to the source).

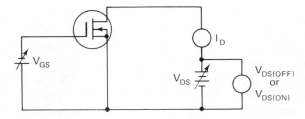

Fig. 6–2 Test circuit for $V_{DS(off)}$ and $V_{DS(on)}$.

6–1.4 I_{DSS} Test

The basic circuit for I_{DSS} test is shown in Fig. 6–3. This test applies to depletion only and depletion/enhancement mode devices. As shown, V_{DS} is set at some fixed value, and V_{GS} is zero (gate shorted to source). I_{DSS} is a zero-bias current flow test. For example, the 3N128 specifies a V_{DS} of 15V, and V_{GS} of zero. Under these conditions, the I_D should be between 5 and 25 mA.

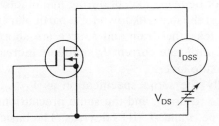

*Disconnect substrate from source (where possible) for $V_{(BR)DSS}$ test.

Fig. 6–3 Test circuit for I_{DSS} and $V_{(BR)DSS}$.

6–1.5 $I_{D(on)}$ Test

$I_{D(on)}$ is essentially the same specification as I_{DSS}, except that $I_{D(on)}$ is generally applied to enhancement mode devices (type B or C). The basic circuit of Fig. 6–1 is used, except that V_{GS} is connected for forward bias (gate positive for N-channel). Forward bias is required since I_D flows in an enhancement mode device only when the gate is forward biased. Some I_D will flow in a class B device without forward bias. However, $I_{D(on)}$ is the current near the maximum (or saturation current). During test, V_{GS} and V_{DS} are set to some fixed value, and the resulting value of $I_{D(on)}$ is measured.

6–1.6 $I_{D(off)}$ Test

The test circuit for $I_{D(off)}$ is the same as for $V_{GS(off)}$, Fig. 6–1. This is because $I_{D(off)}$ is the current that flows when $V_{GS(off)}$ is applied. V_{DS} and V_{GS} are set to some fixed value, and the resulting value of $I_{D(off)}$ is measured. For example, the 3N128 specifies a V_{DS} of 20V, and a V_{GS} of -8V. Under these conditions, the I_D should be 50 μA maximum. On enhancement only mode devices (type C) the value of $I_{D(off)}$ is that I_D that flows when $V_{GS(th)}$ is applied. Thus, the fixed value of V_{GS} is always some forward bias voltage.

6–1.7 Voltage Breakdown Tests

$V_{(BR)GSS}$ is the breakdown voltage from gate to source, and is a *maximum voltage, not a test voltage*. That is, $V_{(BR)GSS}$ is the voltage at which the gate oxide layer breaks down. If this full voltage is applied during test, the oxide layer can (and probably will) be damaged. If the test is performed on a MOS device with gate protection, the resultant breakdown voltage is that of the protective diode, not the MOS device. Thus, $V_{(BR)GSS}$ tests should not be made except during manufacture (as a part of destructive testing to establish breakdown for a particular type of device). To perform a $V_{(BR)GSS}$ test, the drain and source are shorted, and an increasing V_{GS} is applied until gate current I_G suddenly increases, indicating puncture of the gate oxide layer.

$V_{(BR)DGO}$ is essentially the same specification as $V_{(BR)GSS}$, except that breakdown is from gate to drain, and the same precautions apply.

$V_{(BR)DSX}$ applies to type A and B MOS devices. The basic circuit of Fig. 6–1 can be used. During test, reverse bias V_{GS} is applied until the device is cutoff. Usually, the $V_{GS(off)}$ value is used. Then, V_{DS} is increased until there is some heavy flow of I_D.

$V_{(BR)DSS}$ applies to type C MOS devices. The basic circuit of Fig. 6–3 can be used, *except that the substrate must be disconnected from the source* (floating substrate). Of course, the substrate cannot be disconnected on those MOS devices with internal connections between substrate and source, or other element. For type C devices, the gate and source are shorted, since no reverse bias V_{GS} is necessary for cutoff. With V_{GS} at zero, V_{DS} is increased until there is some heavy flow of I_D.

6–1.8 I_{GSS} Test

The basic circuit for I_{GSS} test is shown in Fig. 6–4. This test applies to all modes of operation. As shown, V_{DS} is zero (drain shorted to source), V_{GS} is set to some given (reverse bias) value, and the resultant value of leakage current I_G is measured. For example, the 3N128 specifies a V_{DS} of zero, and a V_{GS} of -8V. Under these conditions I_{GSS} should not exceed 50 pA (at 25°C) or 5 nA (at 125°C).

6–1.9 $V_{G1S(off)}$ and $V_{G2S(off)}$ Test

These tests apply to dual-gate devices, operating in the depletion and depletion/enhancement modes. Dual-gate devices can also be tested for $V_{GS(off)}$ as described in Sec. 6–1.1, when both gates are tied together. The basic circuit for $V_{G1S(off)}$ and $V_{G2S(off)}$ is shown in Fig. 6–5. As shown, V_{DS} is set at some specific value. One gate is forward-biased to a specific value, and the other gate is reverse-biased. The reverse-bias V_{G1S}

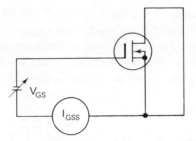

Fig. 6-4 Test circuit for I_{GSS}.

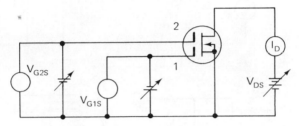

Fig. 6-5 Test circuit for $V_{G1S(off)}$ and $V_{G2S(off)}$.

or V_{G2S} is adjusted until I_D is at some negligible value, indicating a cutoff condition. For example, to measure $V_{G1S(off)}$, the 40841 specifies a V_{DS} of 15V, a V_{G2S} of +4V, and an I_D of 50 μA. V_{G1S} should be −2V for the 50 μA of I_D. To measure $V_{G2S(off)}$, the 40841 specifies a V_{DS} of 15V, V_{G1S} of 0V, and an I_D of 50 μA. V_{G2S} should also be −2V for the 50 μA of I_D.

6-1.10 Dual-Gate Voltage Breakdown Tests

Dual-gate MOS devices are often tested for forward and reverse gate-to-source breakdown voltage. The basic test circuit is shown in Fig. 6-6. As shown, V_{DS} and one gate are at 0V (both shorted to source). A variable voltage is applied to the opposite gate, and the gate current is measured. For example, to measure $V_{(BR)G1SSF}$, the 40841 specifies a V_{DS} and V_{G2S} of 0V, and I_{G1SSF} of 100 μA, with a typical 9V of V_{G1} applied. $V_{(BR)G1SSR}$ is measured in the same way, except that the gate is reverse-biased. The results should be the same (100 μA of I_{G1SSR} for the 40841). $V_{(BR)G2SSF}$ and $V_{(BR)G2SSR}$ are measured in the same way, except that gate 1 is connected to the source, and the voltage is applied to gate 2.

6-1.11 Dual-Gate Current Tests

Dual-gate current tests use the same basic test circuit as gate voltage breakdown tests (Fig. 6-6). The difference in procedure is that the gate voltage is set to a specific value, and the resultant current is measured. For example, to measure I_{G1SSF}, the 40841 specifies a V_{DS} and V_{G2S} of 0V, and a V_{G1S} of 6V. A *maximum* I_{G1SSF} of 60 nA should flow under these conditions. I_{G1SSR} is measured the same way, except that V_{G1S} is −6V. I_{G2SSF} and I_{G2SSR} are measured in the same way, except that gate 1 is connected to the source, and the voltage is applied to gate 2.

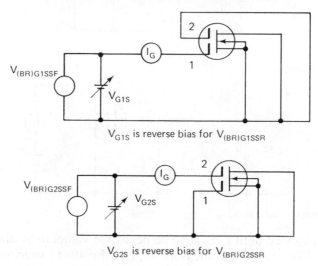

Fig. 6–6 Test circuits for dual-gate voltage breakdown and current.

6–1.12 I_{DS} Test

This test applies to dual-gate devices, operating in the depletion and depletion/enhancement modes. Dual-gate devices can also be tested for I_{DSS} as described in Sec. 6–1.4, when both gates are tied together. The basic circuit for I_{DS} is shown in Fig. 6–7. As shown, V_{DS} is set at some specific value, gate 1 is shorted to the source, and gate 2 is set to a specific value. I_{DS} is considered as a zero-bias current flow test, even though one gate has a forward bias. For example, the 40841 specifies a V_{DS} of 15V, a V_{G1S} of 0V, and a V_{G2S} of +4V. Under these conditions, the I_D should be a typical 10 mA.

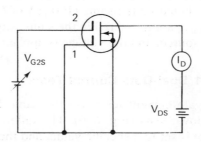

Fig. 6–7 Test circuit for I_{DS} (dual-gate zero bias current test).

6-2. DYNAMIC TESTS

The following paragraphs describe important dynamic (alternating-current) tests for MOS devices. All of the characteristics covered in Sec. 1–6.2 are discussed.

6-2.1 *y*-Parameter Tests

The *y*-parameter tests establish the admittances (forward transadmittance, reverse transadmittance, input admittance and output admittance) of a MOS device. Admittance is the reciprocal of impedance, and is composed of conductance (g, the real part of admittance) and susceptance (b, the imaginary part of admittance). The *y*-parameters can be expressed as complex numbers, where both the g and b values are given, or as a simple number in mhos. This is the same as expressing impedance (Z) in simple terms of ohms, or as a complex number composed of resistance (R) and reactance (X).

When it is necessary to test a MOS device to find a complex number (both the conductance and susceptance), it is necessary to use an *admittance meter* or possibly an R_X meter. An admittance meter consists essentially of a signal source and a bridge circuit. The *y*-parameter under test is connected to form one leg of the bridge. The signal source is adjusted to the frequency of interest, and applied to the bridge. The bridge conductance and susceptance leg elements (usually a resistor and capacitor) are adjusted until the bridge is balanced. The conductance and susceptance values required to produce balance are equal to the *y*-parameter conductance and susceptance at that frequency. The R_X meter functions in essentially the same way, except that the readout is in resistance and reactance. The reciprocal of these R and X values are taken to find g and b.

Operation of admittance meters and R_X meters is discussed in the author's *Handbook of Electronic Test Equipment,* and will not be repeated here. Instead, we will present test procedures that can be accomplished with basic test equipment to find *y*-parameters, expressed in *simple number terms.*

6-2.2 Forward Transadmittance (Transconductance) Test

Forward transadmittance Y_{fs} is often referred to as transconductance, and may be given as y_{21}, g_m, or even g_{fs}.

The basic circuit for Y_{fs} test is shown in Fig. 6–8. This circuit applies to single-gate MOS devices. The circuit for dual-gate is the same, except that a forward bias is generally applied to gate 2. For example, the 40841 specifies a $V_{G2S\ of}$ +4V for test of forward transconductance (which the datasheet lists as g_{fs}). This forward bias produces an I_D of 10 mA when the V_{DS} is 15V. As indicated in Fig. 6–8, the value of R_L must be such that the drop is negligible at I_{DSS}. Just as important, the value of R_L must be such that the V_{DS} will be correct for a given V_{DD} and I_D. For example, if I_D is 10 mA, V_{DD} is 20V, and V_{DS} must be 15V, R_L must drop 5V at 10 mA. Thus, the value for R_L is: 5V/0.01A = 500 ohms. For single-gate devices, the value of I_{DSS} should be used to set the value of R_L. The value of R_G is not critical, and is typically 1 megohm.

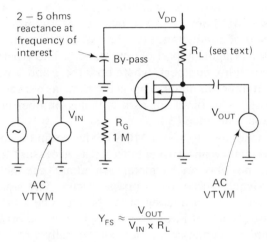

$$Y_{FS} \approx \frac{V_{OUT}}{V_{IN} \times R_L}$$

Fig. 6–8 Test circuit for Y_{fs} (as a simple number).

During test, the signal source is adjusted to the frequency of interest. The amplitude of the signal source V_{in} is set to some convenient number such as 1V, 100 mV, etc. The value of Y_{fs} is calculated from the equation of Fig. 6–8, and is expressed in mhos (or millimhos, or μmhos).

As an example, assume that the value of R_L is 1000 ohms, V_{in} is 1V, and V_{out} is 8V. The value of Y_{fs} is:

$$\frac{8}{1 \times 1000} = 0.008 \text{ mho} = 8 \text{ mmhos} = 8000 \ \mu\text{mhos}$$

6–2.3 *Re(Y_{fs})* Tests

For certain high frequency applications, it is necessary to know the real part of Y_{fs}. At low frequencies, the Y_{fs} value found as described in Sec. 6–2.2 is sufficiently accurate. As frequency increases, Y_{fs} will appear to be *exceptionally high,* if tested by the circuit of Fig. 6–8. To get a true picture of the forward transconductance characteristic, it may be necessary to measure Y_{fs} as a complex number, using an admittance meter or R_X meter. This is almost always true at 30 MHz or higher.

6-2.4 Reverse Transadmittance Test

Reverse transadmittance Y_{rs} (also known as reverse transconductance, or Y_{12}) is generally not a critical value. For practical test purposes, the real value of Y_{rs} (or g_{rs}) can be considered as nearly zero at all useful frequencies. If it is necessary to establish the imaginary part of Y_{rs} (b_{rs}), such as for calculations involved in RF matching networks, an admittance meter or R_X meter must be used.

6-2.5 Output Admittance Test

Output admittance Y_{os} may be given as Y_{22}, g_{os}, or g_{22}. The basic circuit for Y_{os} test is shown in Fig. 6–9. The circuit for dual-gate is the same, except that a forward bias is generally applied to gate 2. As indicated, the value of R_S must be such as to cause a negligible drop (so that V_{DS} can be maintained at a desired level, with a given V_{DD} and I_D).

During test, the signal source is adjusted to the frequency of interest. Both V_{out} and V_{DS} are measured, and the value of Y_{os} is calculated from the equation of Fig. 6–9.

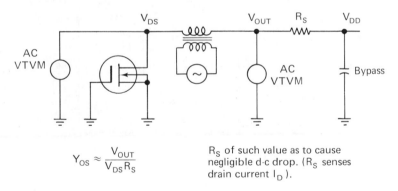

$$Y_{OS} \approx \frac{V_{OUT}}{V_{DS} R_S}$$

R_S of such value as to cause negligible d-c drop. (R_S senses drain current I_D).

Fig. 6-9 Test circuit for Y_{os} (as a simple number).

6-2.6 Input Admittance Test

Input admittance Y_{is} (or Y_{11}) is generally not a critical value. For practical test purposes, the real value of Y_{is} (or g_{is}) can be considered as nearly zero at all useful frequencies. If it is necessary to establish the imaginary part of Y_{is} (or b_{is}), such as for calculations involved in RF matching networks, an admittance meter or R_X meter must be used.

6–2.7 Input Capacitance Test

The basic circuits for C_{iss} test are shown in Fig. 6–10. For dual-gate, the capacitance measurement is between gate 1 and all other terminals. For single-gate, the capacitance measurement is between the gate and all other terminals. The circuit of Fig. 6–10(a) is for MOS devices where a specific V_{GS} must be applied, but V_{DS} is of no concern. The circuit of Fig. 6–10(b) is used where the capacitance is measured by means of a bridge with three terminals (high, low, and guard or ground). The reason for the two circuits is that some MOS device datasheets specify a given V_{DS}, I_D, and V_{GS} when C_{iss} is measured. Other datasheets specify zero V_{DS}, and sometimes zero V_{GS}.

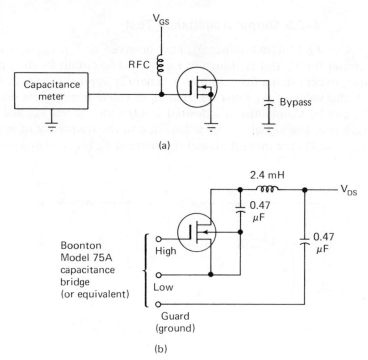

Fig. 6–10 Test circuit for C_{iss}.

6–2.8 Output Capacitance Test

The basic circuit for C_{oss} test is shown in Fig. 6–11. Generally, C_{oss} is used only in dual-gate MOS device specifications. In a typical specification, gate 1 is shorted to the source, and gate 2 has a specific voltage value applied to produce a given I_D.

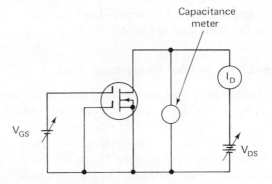

Fig. 6-11 Test circuit for C_{oss}.

6-2.9 Reverse Transfer Capacitance Test

The basic circuit for C_{rss} test is illustrated in Fig. 6-12. A three-terminal measurement is used. For dual-gate, gate 2 and the source are returned to the guard terminal, and the capacitance measurement is made between gate 1 and the drain. Note that the 0.47 μF capacitor in series with the gate-drain connection will be included in the measurement. However, the value of C_{rss} is so low (typically less than 1 pF) that the series capacitance value indicated by the bridge will be that of the MOS device.

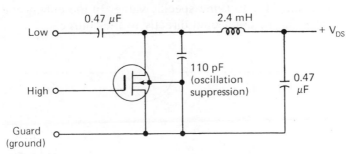

Fig. 6-12 Test circuit for C_{rss}.

6-2.10 Element Capacitance Tests

The capacitance between elements of a MOS device can be measured with a capacitance meter. No special test connections are required. However, some datasheets specify certain test connections or conditions, such as all remaining elements connected to source, or gate connected to source, etc. Generally, the only element capacitances of any consequence are $C_{d(sub)}$ drain-substrate junction capacitance, and C_{ds}

drain-to-source capacitance. Likewise, these capacitances are of little concern, except for switching applications.

6-2.11 Channel Resistance Tests

The basic circuit for channel resistance tests is shown in Fig. 6–13. Both $r_{ds(on)}$ and $r_{ds(off)}$ can be measured with the basic circuit. However, $r_{ds(on)}$ is generally the test of interest.

If the MOS device operates in the depletion or depletion/enhancement modes, $r_{ds(on)}$ is measured by adjusting V_{GS} to 0V (or simply connecting the gate to the source). For example, the 3N128 specifies a V_{DS} and V_{GS} of 0V for an $r_{ds(on)}$ of 200 ohms (at a test frequency of 1 kHz).

During test, the ac voltage source V is adjusted to some convenient value (1V, 10V, etc.), and the channel current I is measured. The value of r_{ds} is found when V is divided by I. For example, if V is adjusted to 10V, and 50 mA channel current is measured, the on channel resistance, or $r_{ds(on)}$, is 200 ohms $(10/0.05 = 200)$.

If the MOS device operates in the enhancement only mode (type C), it is necessary to forward bias the gate by adjusting V_{GS} to some specific value.

Figure 6–13 can also be used to measure the off channel resistance $r_{ds(off)}$. However, the bias conditions are opposite those for $r_{ds(on)}$. In the depletion and depletion/enhancement modes, the gate must be reverse-biased by adjusting V_{GS} to some specific value. In the enhancement only mode, the gate can be connected directly to the source.

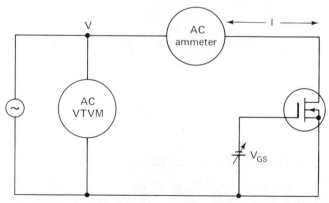

$V_{GS} = 0$ for depletion mode
V_{GS} is forward biased for enhancement mode

$$r_{ds(on)} = \frac{V}{I}$$

Fig. 6–13 Test circuit for $r_{ds(on)}$ and $r_{ds(off)}$.

Figure 6–13 can also be used to measure channel resistance of dual-gate MOS devices. Generally, the simplest way is to connect both gates together. However, some datasheets specify a fixed bias on one or both gates.

6–2.12 Switching Time Tests

Switching time tests for MOS devices are the same as for similar devices (two-junction transistors, etc.). The procedures for finding delay, storage, rise and fall times are discussed fully in the publications described in the introduction to this chapter. A typical test circuit for measurement of MOSFET switching characteristics is shown in Fig. 6–14.

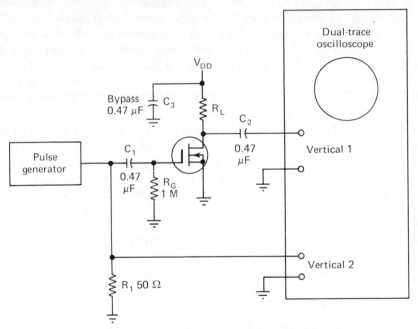

R_L = a value that will cause negligible d-c drop at I_{DSS}
All remaining circuit values are typical.

Fig. 6–14 Test circuit for switching characteristics

6–2.13 Gain Tests

The Y_{fs} test described in Sec. 6–2.2 can be used to establish the gain of a MOS device (MOSFET). However, the test will not establish the gain in a working circuit. The only true test of circuit gain is to operate the device in a working circuit, and measure actual gain. The most practical method is to operate the MOS device in the circuit with which the device is to be used. However, it may be convenient to have a *standard* or *universal* circuit for test of MOSFET gain.

Generally, the main concern is with power gain in RF circuits. Most MOSFET datasheets include a power gain test circuit diagram. Figures 6–15 and 6–16 are typical examples of such power gain test circuits. The circuit of Fig. 6–16 is for dual-gate, where gate 2 is connected to a variable AGC voltage. This permits the power gain to be measured under various AGC conditions. For example, the 40821 specifies power gain between 14 and 17 dB, with V_{DS} of 15V, I_D of 10 mA, and V_{G2S} of $+4$V (applied at the AGC line). Keep in mind that these test circuits are for specific MOSFETs operating with specific loads (50 ohms), and at specific frequencies (200 MHz). However, similar test circuits can be fabricated (using the data of Chap. 3) for other loads and frequencies.

The procedures for measuring power gain are discussed in the publications described earlier. In brief, an RF signal voltage is applied across the input load, and the input power calculated from voltage and load resistance. The amplified output voltage is measured across a similar load, and the power calculated. The ratio of output power to input power is the power gain, and is expressed in dB.

For those MOS devices used as converters, conversion gain is of importance. Figures 6–17 and 6–18 are typical examples of conversion gain test circuits (supplied as part of the datasheet). The circuit of Fig. 6–17 is for a single-gate device where both the RF input and local oscillator input are applied to the gate. Figure 6–18 is for dual-gate where RF input is applied to gate 1, and the local oscillator input is applied to gate 2. Again, these circuits are for specific devices and frequencies. However, the circuits will serve as a design "starting point" for standard test circuits.

6–2.14 Noise Figure Tests

The circuits of Fig. 6–15 and 6–16 can also be used to measure noise figure. Such noise figure tests are best accomplished using specialized test equipment, such as a VHF noise source and VHF noise meter. The instructions supplied with the noise measurement test equipment are generally complete, and provide full descriptions of noise tests. In any event, they will not be duplicated here. However, the test equipment instructions generally omit the amplifier test circuit, thus creating a need for circuits as shown in Figs. 6–15 and 6–16.

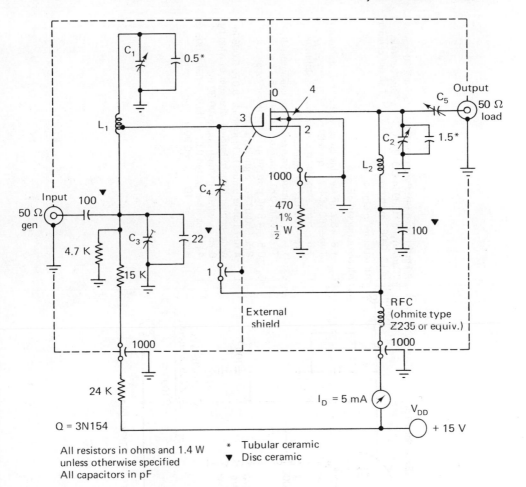

All resistors in ohms and 1.4 W
unless otherwise specified
All capacitors in pF

* Tubular ceramic
▼ Disc ceramic

C_1, C_2: 1.5-5 pF variable air capacitor: E. F. Johnson Type 160-102
or equivalent
C_3: 1-10 pF piston-type variable air capacitor: JFD Type VAM-010,
Johanson Type 4335, or equivalent
C_4, C_5: 0.3-3 pF piston-type variable air capacitor: Roanwell Type
MH-13 or equivalent

L_1: 5 turns silver-plated 0.02″ thick, 0.07″-0.08″ wide copper
ribbon. Internal diameter of winding = 0.25″; winding
length approx. 0.65″. Tapped at $1\frac{1}{2}$ turns from C_1 end
of winding
L_2: Same as L_1 except winding length approx. 0.7″; no tap

Fig. 6–15 Power gain and noise figure (NF) test circuit for 3N128. (Courtesy RCA)

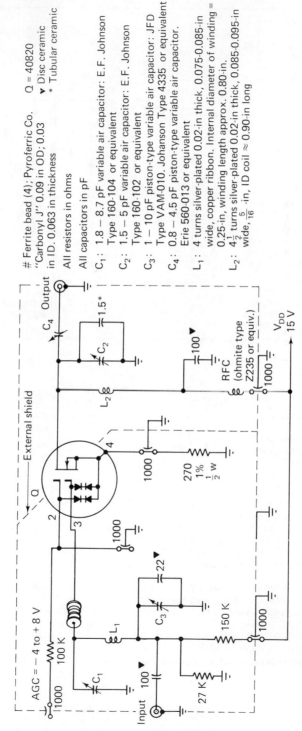

Ferrite bead (4); Pyroferric Co.
"Carbonyl J" 0.09 in OD; 0.03
in ID. 0.063 in thickness

▼ Disc ceramic
Q = 40820
* Tubular ceramic

All resistors in ohms

All capacitors in pF

C_1: 1.8 – 8.7 pF variable air capacitor: E.F. Johnson
Type 160-104 or equivalent

C_2: 1.5 – 5 pF variable air capacitor: E.F. Johnson
Type 160-102 or equivalent

C_3: 1 – 10 pF piston-type variable air capacitor: JFD
Type VAM-010. Johanson Type 4335 or equivalent

C_4: 0.8 – 4.5 pF piston-type variable air capacitor.
Erie 560-013 or equivalent

L_1: 4 turns silver-plated 0.02-in thick, 0.075-0.085-in
wide, copper ribbon. Internal diameter of winding =
0.25-in, winding length approx. 0.80-in.

L_2: $4\frac{1}{2}$ turns silver-plated 0.02-in thick, 0.085-0.095-in
wide, $\frac{5}{16}$ -in, ID coil ≈ 0.90-in long

Fig. 6-16 Power gain and noise figure (NF) test circuit for 40829. (Courtesy RCA)

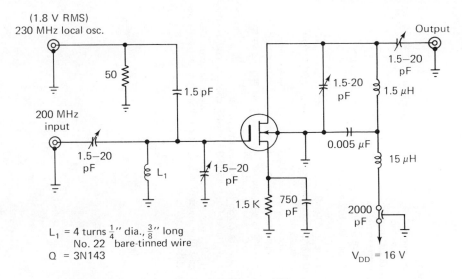

(1.8 V RMS)
230 MHz local osc.

Fig. 6-17 Conversion power gain test circuit for 3N143. (Courtesy RCA)

6-2.15 Cross-Modulation Tests

Figure 6–19 shows the basic block diagram and amplifier circuits for cross-modulation tests. These circuits and values are for a 3N128 being tested at frequencies of 200 MHz (desired frequency) and 150 MHz (interfering frequency). The circuits provide for unneutralized, neutralized, and cascode amplifier operation.

During test, signals of both frequencies are applied to the input. The test circuit output is monitored by the receiver. The amount of attenuation produced on the interfering signal is measured at various signal levels. Typical test results are shown in Figs. 3–25 and 3–26.

6-2.16 Intermodulation Tests

Figure 6–20 shows the basic circuit for intermodulation tests. This circuit and values are for a 3N128 being tested at frequencies of 175 MHz (f_1) and 200 MHz (f_2). If intermodulation is present, there will be an intermodulation signal of 150 MHz, as discussed in Sec. 3–2.6.

During test, f_1 and f_2 are set to zero (amplitude) and the background noise level is measured (on the RF indicator of the receiver at the output of the test circuit). The amplitudes of f_1 and f_2 are increased until the reading on the RF indicator is 1 mV above the noise level. The voltage levels required to produce this output indication are measured on the RF voltmeter at the test circuit input. Typical test results are shown in Fig. 3–27.

Q = 40821
▼ Disc ceramic
* Tubular ceramic
All resistors in ohms
All capacitors in pF

C_1, C_2: 1.5–5 pF variable air capacitor: E.F. Johnson
   Type 160-102 or equivalent
C_3: 1-10 pF piston-type variable air capacitor: JFD
   Type VAM-010, Johanson Type 4335 or equivalent
C_4: 0.9–7 pF compression-type capacitor: ARCO 400
   or equivalent
L_1: 5 turns silver-plated 0.02'' thick, 0.07''-0.08'' wide,
   copper ribbon. Internal diameter of winding = 0.25'',
   winding length approx. 0.65''. Tapped at $1-1\frac{1}{2}$ turns
   from C_1 end of winding
L_2: Ohmite Z-235 RF choke or equivalent
L_3: J. W. Miller Co. # 4580 0.1 μH RF choke or equivalent

Note: If 50 Ω meter is used in place of sweep detector, a low
   pass filter must be provided to eliminate local oscillator
   voltage from load

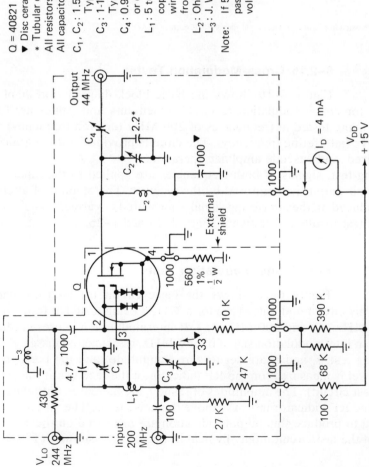

Fig. 6–18 Conversion power gain test circuit for 40821. (Courtesy RCA)

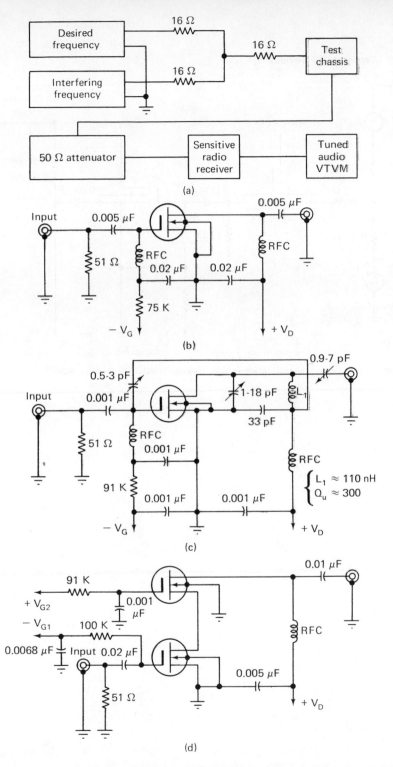

Fig. 6–19 Test circuits used to measure cross-modulation distortion in MOSFETs: (a) block diagram; (b) unneutralized-stage test circuit; (c) neutralized-stage test circuit; (d) cascode-stage test circuit. (Courtesy RCA)

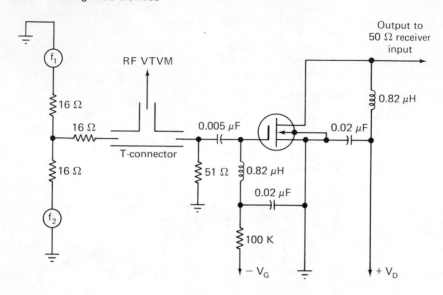

Fig. 6–20 Test circuit used to measure intermodulation distortion in MOSFETs. (Courtesy RCA)

INDEX

A

B